HANDBOOK OF THE MULTIPLICATION OF PLANTS

HANDBOOK OF THE MULTIPLICATION OF PLANTS

By
L.H. Bailey

DISCOVERY PUBLISHING HOUSE PVT. LTD.
NEW DELHI-110 002

Edition – 2016

ISBN: 978-81-7141-488-8

Handbook of the Multiplication of Plants

Published by:
DISCOVERY PUBLISHING HOUSE PVT. LTD.
4383/4B, Ansari Road, Darya Ganj
New Delhi-110 002 (India)
Phone: +91-11-23279245, 43596064-65
Fax: +91-11-23253475
E-mail: discoverypublishinghouse@gmail.com
sales@discoverypublishinggroup.com
web: www.discoverypublishinggroup.com

Printed at:
Infinity Imaging Systems
Delhi

PREFACE TO FIRST EDITION.

THIS little handbook aims at nothing more than an account of the methods commonly employed in the propagation and crossing of plants, and its province does not extend, therefore, to the discussion of any of the ultimate results or influences of these methods. All such questions as those relating to the formation of buds, the reciprocal influences of cion and stock, comparative advantages of whole and piece roots, and the results of pollination, do not belong here.

In its preparation I have consulted freely all the best literature of the subject, and I have been aided by many persons. The entire volume has been read by skilled propagators, so that even all such directions as are commonly recommended in other countries have also been sanctioned, if admitted, as best for this. In the propagation of trees and shrubs and other hardy ornamentals, I have had the advice of the head propagator of one of the largest nurseries in this country. The whole volume has also passed through the hands of B. M. Watson, of the Bussey Institution of Harvard University, a teacher of unusual skill and experience in this direction, and who has added greatly to the value of the book. The articles upon orchids, and upon most of the different genera of orchids in the Nursery List, have been contributed by W. J. Bean, of the Royal Gardens, Kew, who is well known as an orchid specialist. I have drawn freely upon the files of magazines, both domestic and foreign, and I have made particular use of Nicholson's Illustrated Dictionary of Gardening, Vilmorin's Les Fleurs de Pleine Terre, Le Bon Jardinier, and Rümpler's Illustrirtes Gartenbau-Lexikon.

It is believed that the Nursery List contains all the plants which are ordinarily grown by horticulturists in this country, either for food or ornament. But in order to give some clue to the propagation of any which are omitted, an ordinal index has been added, by which one can search out plants of a given natural order or family. It cannot be hoped that the book is complete, or that the directions are in every case best for all regions, and any corrections or additions which will be useful in the preparation of a second edition are solicited.

L. H. BAILEY.

ITHACA, N. Y., January 1, 1891.

PREFACE TO THIRD EDITION.

THIS manual was first published in 1891, by the Rural Publishing Company. In 1892, the publishers made a second edition from the same plates; and both editions are exhausted. The book has had no revision or corrections, therefore, until the present time. It has enjoyed a popularity far beyond its merits, and it has, therefore, seemed worth while to fully revise and recast it, and to make it one of the Garden-Craft Series.

In this revision, it has seemed best to give a somewhat full discussion of the too prevalent assumption that graftage is necessarily a devitalizing process, and to analyze the unclassified knowledge respecting the mutual influences of stock and cion, and the respective peculiarities of root-grafted and budded fruit trees. Something has also been said respecting the so-called exhaustion of nursery land, and of various other nursery matters upon which there seems to be much misunderstanding. The Nursery List now comprises the notes and suggestions of many correspondents, and the results of the experience and experiment of five additional years. The entire volume has been thoroughly ransacked and renovated, and in this work I have been aided by B. M. Watson, to whose efficient aid the first edition owed so much, and by my associate, E. G. Lodeman.

The chapter upon pollination has been omitted in this edition, because a similar one has been incorporated in my "Plant-Breeding." The ordinal index, which was a separate feature of the other editions, is now included in the regular index.

L. H. BAILEY.

CORNELL UNIVERSITY,
ITHACA, N. Y., July 1, 1896.

CONTENTS.

CHAPTER I.

PAGE

SEEDAGE 1–25
1. Requisites of Germination 1
 Regulation of Moisture 1
 Requirements of Temperature 7
 Influence of Light upon Germination 8
 Regeneration 9
2. Seed-Testing 9
3. The Handling and Sowing of Seeds and Spores 15
 Preparatory Treatment of Seeds 15
 Transportation of Seeds from Abroad 19
 Sowing 20
 Damping-off 23
 Spores 24

CHAPTER II.

SEPARATION AND DIVISION 26–34
1. Separation 26
2. Division 32

CHAPTER III.

LAYERAGE 35–43

CHAPTER IV.

CUTTAGE 44–72
1. General Requirements of Cuttings 44
 Devices for Regulating Moisture and Heat 44
 Bottom Heat 53

PAGE
Soils 54
The Formation of Roots 55
2. The Various Kinds of Cuttings 58
Tuber Cuttings 59
Root Cuttings 60
Stem Cuttings 62
Leaf Cuttings 70

CHAPTER V.

GRAFTAGE 73–156
1. General Considerations 73
Mutual Influence of Stock and Cion 74
Limits of Graftage 77
General Methods 78
Classification of Graftage 79
Is Graftage a Devitalizing Process? 81
2. Budding 94
Shield-budding 95
Prong-budding 105
Plate-budding 105
H-budding 106
Flute-budding 106
Chip-budding 107
3. Grafting 107
Whip-grafting 108
Modified Whip-grafts 111
Saddle-grafting 113
Splice-grafting 113
Veneer-grafting 113
Side-grafting 115
Inlaying 117
Cleft-grafting 118
Bark-grafting 129
Herbaceous-grafting 130
Seed-grafting 131
Cutting-grafting 131

PAGE
Double-working ... 133
Inarching ... 132
Grafting Waxes ... 134
4. Nursery Management ... 138
Nursery Lands ... 139
Grades of Trees ... 142
The Storing of Trees ... 143
Trimming Trees in the Nursery ... 146
Dwarfing ... 147
Root Grafted vs. Budded Trees ... 148

CHAPTER VI.

THE NURSERY LIST ... 157–336

GLOSSARY ... 337

INDEX ... 349

CHAPTER I.

SEEDAGE.

1. REQUISITES OF GERMINATION.

THERE are three external requisites to the germination of seeds—moisture, free oxygen, and a definite temperature. These requisites are demanded in different degrees and proportions by seeds of different species, or even by seeds of the same species when differing widely in age or in degree of maturity. The supply of oxygen usually regulates itself. It is only necessary that the seeds shall not be planted too deep, that the soil is porous and not overloaded with water. Moisture and temperature, however, must be carefully regulated.

Regulation of Moisture.—Moisture is the most important factor in seedage. It is usually applied to the seeds by means of soil or some similar medium, as moss or cocoanut fiber. Fresh and vigorous seeds endure heavy waterings, but old and poor seeds must be given very little water. If there is reason to suspect that the seeds are weak, water should not be applied to them directly. A favorite method of handling weak and also

1. Double seed-pot.

very small seeds is to sow them in a pot of loose and sandy loam which is set inside a larger pot, the intermediate space being filled with moss, to which, alone, the water is applied. This device is illustrated in Fig. 1. The water soaks through the walls of the inner pot and is supplied gradually and constantly to the soil. Even in this case it is necessary to prevent soaking the moss too thoroughly, especially with very weak seeds. When many pots are required, they may be simply plunged in moss with the same effect. The soil should be simply very slightly moist, never wet. Moisture is sometimes supplied by setting the seed-pot in a shallow saucer of water, or it may be sufficient to simply place it in the humid atmosphere of a propagating-box. Large but weak seeds may be laid upon the surface of the soil in a half-filled pot, covered with thin muslin, and then covered with loose and damp loam. Every day the pot is inverted, the covering taken off and fresh soil added. A modification of this plan, for small seeds, can be made by placing the seeds between two layers of thin muslin and inserting them in damp loam, which is frequently renewed to avoid the extremes which would result from watering or from allowing the soil to become dry. In these last operations, no water is applied to the seeds, and they constitute one of the most satisfactory methods of dealing with seeds of low viability. They are essentially the methods long ago used by Knight, who laid such seeds between two sods cut from an old and dry pasture.

Even sound and strong seeds should be watered with care. Drenchings usually weaken or destroy them. The earth should be kept simply damp. To insure comparative dryness in indoor culture, some loose material, as pieces of broken pots or clinkers, should be placed in the bottom of the pot or box to afford drainage. It should be borne in mind, however, that the seed-bed should be approximately equally moist throughout its depth. The waterings should, therefore, be copious enough to moisten the soil throughout.

A wet or moist surface over a dry substratum should always be avoided. Error is common here. It is usually best to apply water with a watering-pot, as watering with a hose is apt to wash out the seeds and to pack the soil, and the quantity of water is not so easily regulated.

At first thought, it would appear that the apparently good results following soaking of seeds in many cases are a contradiction of these statements that seeds may be over-watered. But soaking is usually beneficial only when practiced for a comparatively short time. It is not good practice to soak delicate seeds before sowing, and it is of doubtful utility in most other cases, unless it is necessary to soften the integuments of hard-shelled species, as discussed on page 16. The gain in rapidity of germination following soaked, as compared with dry seeds, is often fictitious, inasmuch as germination actually begins in the soaked seed before the dry samples are sown. The soaked seeds are sown in water rather than in soil, and as conditions are more uniform there, a gain apparently due to soaking may result. In the case of strong seeds which must be planted outdoors in cold or uncongenial soil, a preliminary soaking of from 12 to 24 hours may be beneficial, as it lessens the period which the seeds would otherwise pass in untoward conditions. But soaked seeds, unless of very hardy species, should never be sown outdoors until the soil has become rather dry and warm.

2. Seed-pot, covered with glass.

To prevent too rapid drying out, the soil should be firmly pressed about the seeds. The pot or box should be given a shady place, or some covering may be applied to check evaporation. A pane of glass is often placed over the pot (Fig. 2) or box, being tilted a little at intervals to allow of ventilation and to prevent the soil from becoming soggy or "sour." A seed case; with a glass cover, as shown in Fig. 3, is neat and handy in the treatment of small seeds.

A thin covering of fine moss is sometimes given, or a newspaper may be thrown over the soil.

In outdoor culture, only a naturally dry and well-drained soil should be chosen for all ordinary seeds, especially for such as are sown in the fall or remain in the ground a long time before germinating. Soils which contain a liberal amount of sand or gravel are especially valuable for this purpose.

3. Glass-covered seed-case.

To prevent drying in outdoor culture, it is important that the earth be well firmed over the seeds. Walking on the row, placing one foot directly ahead of the other, is usually the most expeditious and satisfactory operation, at least with large seeds. Or the earth may be firmed with a hoe or the back of a spade, or a board may be placed upon the row and then be thoroughly settled by walking over it. For small lots of seeds, it is well to cover them with an inverted flower-pot (Fig. 4), exercising care to tilt it frequently to prevent the plants from "drawing." In the sowing of celery and other small and slow seeds, it is a frequent practice to leave the board on the row until the seeds appear, in order to hold the moisture. This is a doubtful expedient, however, for the young plants are apt to be quickly dispatched by the sun when the board is removed. If the board is employed, it should be raised an inch or two from the ground as soon as the plants begin to appear. But the shade of the board is too dense, and plants do not grow stocky under it. It is better to use brush or lath screens if protection is desired; or fine litter, if free from weed seeds, may be used. In most cases, however, screens will not be needed

4. Seeds covered with flower-pot.

by celery and similar seeds if the ground is in the proper condition, so that it will neither bake nor dry out quickly, and is well firmed at planting time, and if the seeds are sown early, before hot, dry weather comes. It is always advisable, nevertheless, to place the beds for slow and small seeds where they can be watered occasionally.

5. Lath screen.

There are many kinds of screens in use to prevent the drying out of small seeds in outdoor seedage and to protect the young seedlings. These are used alsɔ in the shading of cuttings. The common lath screen (Fig. 5) is the most useful for general purposes. It is simply a square frame made from common laths laid at right angles in a double series. The interstices between the laths are equal in width to the laths themselves. These screens are laid horizontally upon a light framework a few inches above the seeds. The passage of the sun constantly moves the shadows over the bed,

6. Brush screen.

and sufficient shade is afforded while thorough ventilation is allowed. This and all other elevated screens are useful in shading and protecting the young plants as well, but when used for this purpose they are usually raised a greater distance above the beds. A brush screen, consisting of a low frame covered with boughs, is often used, as shown in Fig. 6. This is cheaper than the lath screens, and is equally as good

7. Screen for frames.

for most purposes. The brush is often laid directly upon the ground, especially in large beds. This answers the purpose of shading, but it does not allow of weeding, and it must be taken off soon after the seeds germinate, or slender plants will be injured in its removal. Brush screens are sometimes raised three or four feet to allow of weeding. A screen for frames is shown in Fig. 7. It is a simple covering of muslin stretched over the top and sides of a rough framework. The cloth is usually omitted from the front side. This style of screens is much used by nurserymen, especially for cutting-beds. Whitewashing the sashes of coldframes also affords good shading. A more elaborate and permanent screen is shown in Fig. 8. It is built of slats, usually 3-inch stuff. This shed screen is oftenest used for the protection of tender plants, but it affords an exceedingly use-

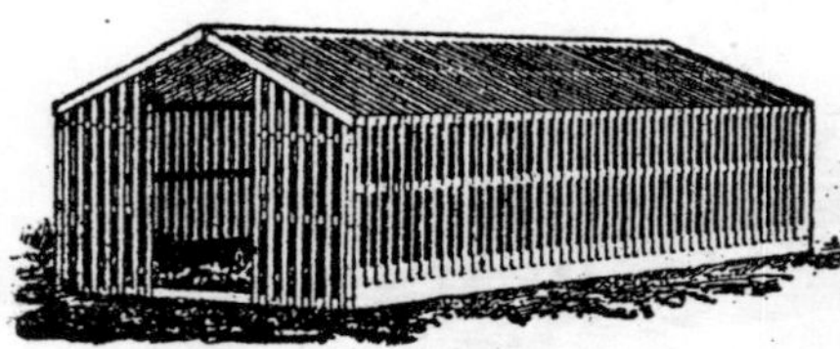

8. Shed screen for seeds and plants.

ful and convenient place for the storage of pots and boxes of slow-germinating seeds. A more elaborate shed screen, made of lath or slats, and containing seed-beds edged with boards, is shown in Fig. 9.

Various frames and covers are employed for indoor seedage, but they are designed to regulate atmospheric moisture and to control temperature. They are more

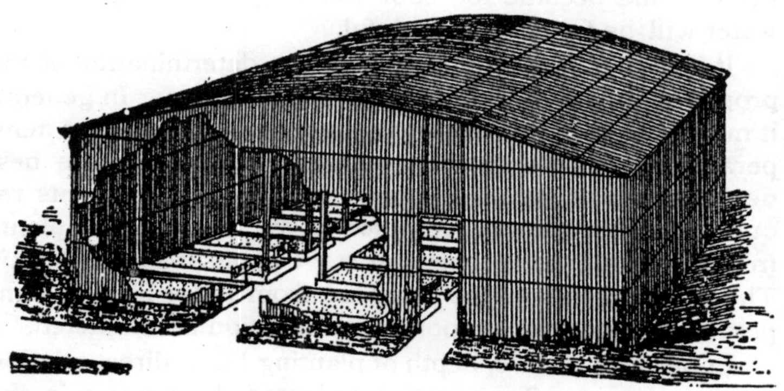

9. Large shed screen, with seed-beds.

commonly employed in the growing of cuttings, and are, therefore, described in Chapter IV.

Requirements of Temperature.—Variations in temperature exercise less influence upon seeds than variations in moisture. Yet it is important that the extremes of temperature should not be great, especially in small, delicate or weak seeds. Seeds will endure greater extremes of temperature when dry than when moist. This indicates that germinating seeds must be kept in a comparatively uniform temperature. For this reason it is poor practice to put seed-boxes in a window in full sunlight. Partial or complete shade serves the double purpose of preventing too great heat and too rapid evaporation. Various covered seed-boxes are used for the purpose of maintaining approximately the required temperature, but as they are oftener used in bud-propagation, they are discussed in that connection.

Bottom heat is helpful to germination in most seeds, but, except in the case of certain tropical species, it should not be strong. It is a common practice to place seed-boxes on moderately cool pipes under benches in a greenhouse. Seeds of hardy annuals and perennials do not require botton heat, although they may be benefited by it. If the soil in seed-beds should become too cool, watering with warm or tepid water will be found to be helpful.

It is impossible to give rules for the determination of the proper temperature for different kinds of seeds. In general, it may be said that seeds germinate most rapidly at a temperature a few degrees above that required for the best development of the plant itself. Seeds of hardy plants require a temperature of from 50° to 70°, conservatory plants from 60° to 80°, and tropical or stove plants from 75° to 95°. The plantlets should be removed from these highest temperatures, as a rule, as soon as germination is completed.

In outdoor culture, depth of planting has a direct relation to temperature. Seeds may be planted deeper late in the season than early, when the soil is cold and damp. Deep planting probably as often kills seeds because of the absence of sufficient heat as from the lack of oxygen or the great depth of earth, through which the plantlet is unable to push.

Influence of Light upon Germination.—The influence which light exerts upon germination is not definitely understood for all horticultural seeds. It is known, however, that seeds will often germinate in full sunlight, if the proper conditions of moisture and temperature can be maintained. Seeds sown upon a moist surface and covered with a glass present an interesting study. But it is well known, on the other hand, that some seeds will not germinate, or will at least appear unevenly, if subjected to sunlight. At least some of the delphiniums, papavers and adonises germinate very imperfectly, if at all, in direct light. It is always advisable to keep germinating seeds in shade or partial darkness, especially as there is nothing to be gained by exposing them. Of course, the soil itself is sufficient protection if the seeds are covered.

Regermination.—It is a common statement that seeds can never revive if allowed to become thoroughly dry after they have begun to sprout. This is an error. Wheat, oats, buckwheat, maize, pea, onion, radish and other seeds have been experimented upon in this direction, and they are found to regerminate readily, even if allowed to become thoroughly dry and brittle after sprouting is well progressed. They will even regerminate several times. Wheat, peas and other seeds have been carried through as many as seven germinations after the radicle had grown a half inch or more and the seeds had been sufficiently dried in each trial to render them fit for grinding.

2. SEED-TESTING.

Whilst it is not the province of this handbook to discuss the question of the testing of seeds, a few hints upon the subject may be acceptable, particularly in the bearing of the remarks upon seed sowing. Germination is complete when the plantlet begins to assume true leaves and to appropriate food directly from the soil. The testing of seeds is not always concerned with germination, but with the simple sprouting of the samples. Many seeds will sprout which are not strong enough to germinate completely, and more seeds will be counted as viable when they are tested in some germinating apparatus—where the conditions are ideal—than when they are normally planted in the soil. There is even sometimes a marked difference between the results of seed-tests made in soil in the greenhouse and in outdoor planting, as the following comparisons (Bulletin 7, Cornell Experiment Station, 1889) plainly show:

"It has been said recently that the ideal test of seeds is actual sowing in the field, inasmuch as the ultimate value of the seed is its capability to produce crop. This notion of seed-tests is obviously fallacious, although the statement upon which it is based is true. In other words, actual planting rarely gives a true measure of the capabilities of all the seeds of any sample, because of the impossibility to control conditions and methods in the field. The object of seed-tests is to determine how many seeds are viable, and what is their relative vigor; if planting shows poorer results, because of covering too deep or too shallow, by exposing to great extremes of temperature or moisture, or a score of other untoward conditions, the sample cannot be held to account for the shortcoming. The following table indicates the extent of variations which may be expected between tests and actual plantings of seeds from the same samples:

"Various samples were tested indoors and actually planted in the field. The seeds were sown in the field June 5, and the last notes were taken from them July 5. They were sown on a gravelly knoll. Rain fell about every alternate day, and the soil was in good condition for germination throughout the month. The indoor tests were made in loose potting earth, or in sand in seed-pans.—

SAMPLES.	No. of germ. in house.	Per cent of germ. in house.	No. of germ. in field (200 seeds sown).	Per cent of germ. in field.	Per cent of difference.
Endive, Green Curled, Thorburn (200 seeds)	88	44	53	26.5	17.5
Tomato, Green Gage, Thorburn (100 seeds)	72	72	93	46.5	25.5
Turnip, Ea. Six Weeks, Dept. of Agriculture (200 seeds)	180	90	65	32.5	57.5
Pea, White Garden Marrowfat, Thorburn (60 seeds)	55	91.6	181	90.5	1.1
Celery, White Plume, Thorburn (100 seeds)	41	41	22	11	30
Onion, Red Wethersfield, Thorburn (200 seeds)	148	74	84	42	32
Carrot, Early Forcing, Thorburn (100 seeds)	70	70	39	19.5	50.5
Carrot, Vermont Butter, Hoskins (100 seeds)	65	65	45	22.5	42.5

"The table indicates that actual planting in the field gives fewer germinations than careful tests in conditions under control. This difference in total of germination, even under favorable conditions of planting, may amount to over 50 per cent.

"In planting, due allowance should be made for the comparatively bungling methods of field practice by the use of greater quantities of seeds than would seem, from the results of tests, to be sufficient."

Probably the most truthful test of seeds can be made in soil in earthen pans in a greenhouse or forcing-house. When one desires to show the ultimate percentage of seeds which contain life, the sprouting test should be used. In this case, some apparatus should be employed in which the moisture and temperature can be controlled to a nicety, and in which the seeds can be examined as often as desired. As soon as a seed sprouts, it is removed and counted as viable, wholly independently of whether it is strong enough to make a plant under ordinary conditions. In other words, the sprouting test is almost wholly an attempt to arrive at a numerical estimate of the viability of the sample, rather than an effort to determine the relative strength of germinative power.

There are many excellent devices for the making of sprouting tests, only three or four of which need be mentioned here, for the purpose of illustrating some of the principles which are employed. One of the best known of these apparatus in this country is the Geneva tester, which originated at the New York Experiment Station at Geneva.

A full account of this device by Professor J. C. Arthur (Botanical Gazette, 1885, p. 425) is here inserted:

"Various methods have been used for testing the per cent and time of seed germination. Those most commonly adopted in this country and also abroad have been to place the seeds on the surface of porous tile, smooth sand or compacted earth. Without stopping to point out the defects and inconveniences of these methods, I desire to describe an apparatus devised at the New York Agricultural Experiment Station, and which has been found so satisfactory as to supersede all other sorts of germinators at that institution for general use. It consists (Fig. 10) of a pan 10 x 14 inches wide and 3½ inches deep, to be covered with a pane of glass. Along the sides is a ledge ⅜-inch wide, and as much below the upper edge. The pan is best made of tinned copper, the ledge formed by the proper shaping of the sides of the pan, and the edges on three sides turned over to form a groove into which the pane of glass may be slid from one end. These details are not shown in the cut. The seeds are held in the folds of cloth. A strip of white Canton flannel is taken sufficiently wide so that when hemmed on both sides (to prevent seeds slipping out of the ends of the folds) it will be the same as the inside width of the pan. A long enough strip is used to have about twenty-four folds 1½ inches deep, and leave a flap of several inches at each end. The upper margin of the folds is sewn across to permit a ⅛-inch brass rod to be run in (Y, P), from which the cloth is suspended in the pan as shown in the cut. The lower margins of the folds (O) are also sewn across to make them stay in place better. The total length of the strip after the sewing is completed is about a yard. Two such strips are used in each pan.

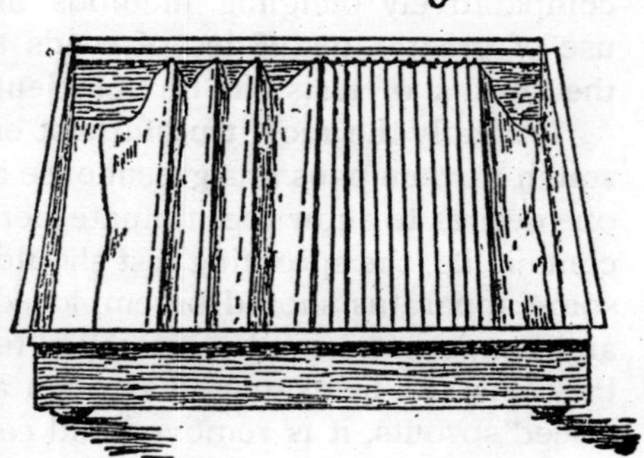

10. The Geneva seed-tester.

"To put the pan into use, it is filled part full of water, two of the prepared cloths put in, the glass cover adjusted and the whole boiled over a lamp for a short time. This is necessary in order both to thoroughly wet the cloth and to kill any mold or other germs. When again cool, adjust the cloths on the brass rods and put in the seeds. Each fold will hold 25 large seeds, like beans, and a hundred or more small seeds. Water is placed in the pan, but not enough to touch the folds of cloth; the four flaps drop down into it, however, and keep the cloths sufficiently wet by capillarity, which is increased by the long nap on the under surface of the cloth. The folds are numbered consecutively, and the record kept by the numbers.

"The advantages in a pan of this kind are the facility with which the seeds may be examined and counted, the thorough and uniform moisture of the seeds throughout the longest trials, its lightness and cleanliness. It is necessary to renew the cloths from time to time, as they will slowly rot out, even with tne best of care."

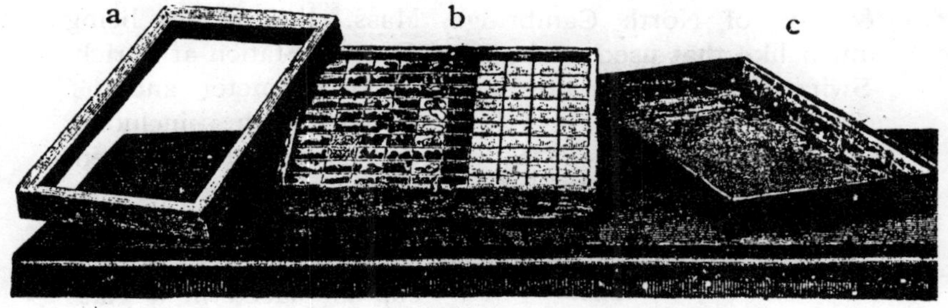

11. An absorbing-block seed-tester.

A device of a wholly different character, used in Germany, is shown in Figs. 11 and 12 (Annals Hort., 1890, 268). It consists of three parts: a tin tray (c) for holding water; a block of gypsum (b) which sits in the tray and contains several compartments for the reception of the seeds, and which is kept moist by capillary attraction; a glass cover (a). The apparatus is seen at work in Fig. 12. This device works upon a principle which has long been utilized in the testing of seeds—the capillary power of earthenware and various species of rock—and there are many applications of the idea in practice. These seed-testers may be placed in an incubator or other heating device, or they may be used in the greenhouse or a living-room.

12. The seed-tester (No. 11) set up.

A recent bulletin (No. 35) of the Rhode Island Experiment Station describes and illustrates a modification of the absorbing-block idea. Instead of a slab of stone or earthenware, "sprouting cups" are used. "These were made for this station out of porous clay by A. H. Hews & Co., of North Cambridge, Mass., the pattern being much like that used at the Seed Control Station at Zurich, Switzerland. They are 3 inches in diameter and 1¾ inches high, including the cover, which is ventilated, as shown in Fig. 13. The bottom is solid and ½-inch thick. Each cup is placed in a glass dish in which a constant supply of water is kept." These cups are placed in a "sprouting chamber" (Fig. 14), supplied with uniform heat. "This holds about fifty cups. Heat is supplied by a gas jet, which is supported beneath the chamber, and it is distributed evenly to all sides of the chamber, except the front, by means of a water-jacket. It is provided with two doors, the inner one being glass. There is an opening in the side and top for ventilating, and a second opening in the top for the insertion of a thermometer. There are also two openings into the water-jacket

13. Sprouting cup.

at the top. In one of these a thermostat (C) is placed, which controls the flow of gas at the jet beneath, and in the other a thermometer (D) may be placed to show the temperature of the water in the jacket."

3. THE HANDLING AND SOWING OF SEEDS AND SPORES.

Preparatory Treatment of Seeds.—Many seeds demand some treatment preparatory to sowing. Nearly all hard and bony seeds fail to germinate, or at least germinate very irregularly, if their contents are allowed to become thoroughly dry and hard. The shells must also be softened or broken, in many cases, before the embryo can grow. Nature treats such seeds by keeping them constantly moist under leaves or mold, and by cracking them with frost. This suggests the practice known to gardeners as *stratification*, an operation which consists in mixing seeds with earth and exposing them to frost or to moisture for a considerable time.

14. Sprouting chamber.

Stratification is practiced, as a rule, with all nuts, the seeds of forest trees, shrubs, the pips of haws and often of roses, and in many cases with the seeds of common fruits. Seeds should be stratified as soon as possible after they are mature. Small seeds are usually placed in thin layers in a box alternating with an inch or two of sand. Sometimes the seeds are mixed indiscriminately in the sand, but unless they are large it is difficult to separate them out at sowing-time. The sand is often sown with the seeds, however, but it is difficult in such cases to distribute the seeds evenly, and in sowing large quantities the handling of the sand entails a considerable burden and becomes an item of expense. It is advisable to pass the sand through a sieve of finer mesh than the seeds, and the seeds can then be sifted out at sowing-time. If the seeds are very small or very few in number, they may be placed between folds of thin muslin, which is then laid in the sand. Any shallow box, like a gardener's "flat," is useful in making stratifications, or pots may be used with small lots of seeds. A flat four inches in depth might contain two or three layers or strata of seeds the size of peas.

The disposition of the boxes when filled varies with different operators. Some prefer to bury them. In this case a well-drained sandy slope is chosen. The flats are placed in a trench from one to two feet deep, covered with a single thickness of boards, and the trench is then filled with earth. The seeds usually freeze somewhat, although freezing is not considered necessary unless in the case of nut-like seeds. The object attained in burying is to keep the seeds moist and fresh, inducing the rotting or softening of the coverings, while they are buried so deep that they will not sprout. Seeds of most forest trees should be treated in this manner. They are commonly left in the ground until the following spring, when they are taken up and sown in drills in mellow ground. If good loam, to which has been added a little well-rotted manure, is used, the seeds or nuts of hardy trees and shrubs may be allowed to germinate and grow for one season in the flats. At the end of the season or the next spring, the plants can be transplanted without losing one. This is, perhaps, the best way to handle rare and difficult subjects.

Many growers place the boxes on the surface in some protected place, as under trees or in a shed, and cover them during winter a foot deep with clean straw or leaves. If boxes are piled on top of each other they should be mulched with moss, else the under ones may become too dry. Or the boxes may be placed, without covering, in a shed, but they must be examined occasionally to see that they do not become too dry. Precaution must also be taken to keep away mice, squirrels, blue-jays, and other intruders.

Large, nut-like seeds or fruits, like peach-pits, walnuts and hickory-nuts, are usually buried in sand or light loam where they may freeze. Or sometimes the large nuts are thrown into a pile with earth and allowed to remain on the surface. Freezing serves a useful purpose in aiding to crack the shells, but it is not essential to subsequent germination, as is commonly supposed. All seeds, so far as known, can be grown without the agency of frost, if properly handled.

Fall sowing amounts to stratification, but unless the soil is mellow and very thoroughly drained the practice is not advisable. The seeds are liable to be heaved or washed out, or eaten by vermin, and the soil is apt to bake over them. Under proper conditions, however, the seeds of fruits and many forest trees thrive well under fall sowing. The seeds should be sown as soon as they are ripe, even if in midsummer; or if the ground is not ready for them at that time, they may be temporarily stratified to prevent too great hardening of the parts. It is best, however, to allow all green or moist seeds to dry off a few days before they are stratified. Fall-sown seeds should always be mulched.

Some seeds rarely germinate until the second year after maturity, even with the best of treatment. The thorns, mountain ash, hollies, viburnums, some roses, and many others belong to this category. Some growers sow them regularly as soon as they are ripe, and allow the beds to remain until the seeds appear. This is a waste of land and of labor in weeding, and the best way is to stratify them and allow them to remain until the first or second spring before sowing.

Partial substitutes for stratification are soaking and scalding the seeds. Soaking may be advantageously practiced in the case of slow and hard seeds which are not enclosed in bony shells, and which have been allowed to become dry. Seeds of apple, locust, and others of similar character, are sometimes treated in this manner. They are soaked for 24 or 36 hours, and it is commonly supposed that if they are exposed to a sharp frost in the meantime, better results will follow. While still wet the seeds are sown. Scalding water may be poured over locust and other seeds to soften their coverings, but seeds should not be boiled, as sometimes recommended.

The germination of bony seeds is often facilitated by filing or cutting away the shell very carefully near the germ, or by

15. Bored seed.

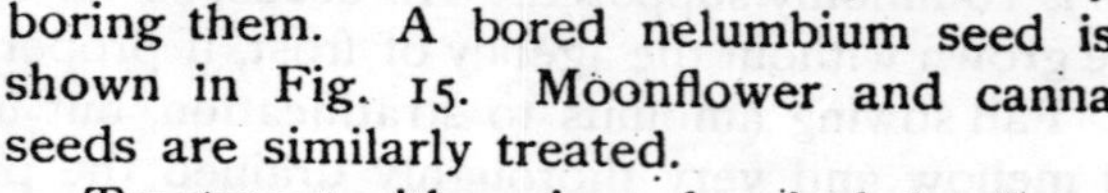

boring them. A bored nelumbium seed is shown in Fig. 15. Moonflower and canna seeds are similarly treated.

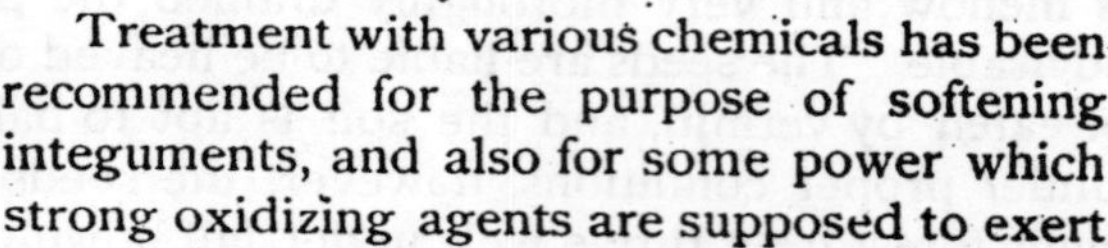

Treatment with various chemicals has been recommended for the purpose of softening integuments, and also for some power which strong oxidizing agents are supposed to exert in hastening germination itself, but the advantages are mostly imaginary. Secret and patented "germinator" compounds had better be avoided.

Pulpy and fleshy coverings should be removed from seeds before sowing. Soft fruits, like berries, are broken up or ground into a pulp, and the seeds are then washed out. This separation may be performed immediately in some cases, but when the pulp adheres to the seed, the whole mass is usually allowed to stand until fermentation and partial decay have liberated the seeds. The pulp will then rise, in most instances, leaving the seeds at the bottom of the vessel. Seeds can be liberated quickly by adding a stick of caustic potash to each pail of water. After the mass has stood an hour or so, the seeds can be rubbed out easily. Even tomato seeds can be cleaned with safety in this manner. Seeds which have thin pulp, as the viburnums and many haws, can be prepared by rubbing them through the hands with sharp sand. Or the scant pulp of such seeds may be allowed to rot off in the stratification box. Fleshy coverings of hard and bony seeds may be removed by maceration. Allow them to stand in water at a temperature of about 75° for one to three weeks, and then wash them out. Resinous coverings are sometimes removed by mixing the seeds with fresh ashes or lime, or by treating them with lye. Hard, thick-walled seeds are rarely injured by the decay of the pulpy covering, but thin-walled seeds should be cleaned, to avoid the possibility of damage arising from the decay of the pulp.

Transportation of Seeds from Abroad.—The transportation of certain kinds of seeds over long distances, especially on sea voyages, is often beset with difficulties. Thick-meated or soft seeds may become too dry if stored in a warm place or too moist if stored in a cool one. The humid atmosphere of the ocean is fatal to some seeds unless they are well protected, and the moist and hot climates of some tropical countries destroy many seeds of cooler regions before they can be planted, or cause them to sprout in transit. Thin-coated seeds demand dryness and air, and bony seeds usually need moisture and a more confined atmosphere. Most seeds may be sent dry and loose in coarse paper packages under all ordinary circumstances; but if they are to traverse very hot and moist climates, they should be sealed in tin cases or very securely wrapped in oiled paper, in which case the seeds should be thoroughly dried before being packed, and precautions taken to insure the dryness of the air in the package. Small seeds which are liable to become moldy may be packed in finely powdered charcoal. Apple and pear seeds are often imported in this manner. The seeds or fruits of woody plants require more careful management. They should generally be transported in some sort of stratification. A favorite method is to place them in boxes or jars, mixed with naturally moist sand or sawdust, or slightly moist dead sphagnum moss. Some prefer to seal the packages hermetically, but under ordinary conditions this is unnecessary. In transit, the packages should be stored in a medium and uniform temperature. Even acorns, which are often difficult to transport over long voyages, may be carried in this manner with safety. It is important that the soil should not be wet. Natural soil from a dryish and loamy pasture is excellent. In some cases it is better to sprout the seeds in the native country and ship the seedlings in a closed or Wardian case.

Sowing.—The soil in which seeds are sown, especially in indoor culture, should be such as to allow of perfect drainage and at the same time to hold moisture. Good potting soil, with a liberal allowance of sharp sand, is the best for general purposes. Pure sand becomes too dense, and leaf mold alone is usually too loose and open. A proper combination of the two corrects both faults. It is impossible to describe a good potting or seed-bed soil. Some experience is essential to the best results in preparing it. It should be of such character that when a damp portion is firmly compressed in the hand it will fall apart when released. It should never bake. Good old garden loam, to which an equal quantity of sand has been added, is usually a good soil for common indoor seedage. There should be no manure in soil used for seeds which produce a delicate growth, as rhododendrons and kalmias. In all such cases, rotted sod or leafy peat is an excellent medium. Live sphagnum moss is also a good material upon which to sow various heath-like seeds, as kalmias, andromedas, and the like. Soil should be sifted and thoroughly fined before seeds are put into it. Seeds usually require lighter soil than that in which the growing plant will flourish. Cocoanut fiber is sometimes used in place of the soil, as it holds moisture, allows of almost perfect drainage, and does not become "sour." Fine dead sphagnum moss may also be used. Orchid seeds are usually sown on the live moss in which the parent plant is growing; or they may be sown on damp wood or cork. (See under Orchids, Chap. VI.) Small seeds, like those of cineraria and calceolaria, germinate well in very old cow-dung obtained from a pasture, from which the unctuous matters have disappeared, leaving a fibrous remainder. But all things considered, well-prepared soil is the most satisfactory medium which can be used for most seeds.

Seeds of aquatic plants, which are to be sown in a pond, may be placed in a ball of clay and dropped into the water. Water lily seeds may be sown in the greenhouse in submerged pots or pans.

Shallow boxes or "flats" and earthen seed-pans and lily-pans are usually preferable to pots in which to sow seeds. They give more surface in proportion to their contents, and require less attention to drainage. If pots are used, the 4 to 6-inch sizes are best. All delicate seeds, like tuberous begonias, primulas, gloxinias, and also spores, are generally sown in pots or pans, which are covered with a pane of glass. (See Figs. 2 and 3.)

If delicate seeds are sown outdoors, they should be given some protection, if possible. An ordinary hotbed frame gives the best results. In warm weather or a sunny exposure it will be found desirable to substitute a cloth screen for the sash. A thin or medium water-proof plant-cloth, either commercial or home-made, is excellent for this purpose. It may be tacked upon a simple and light rectangular frame which is strengthened at the corners by iron "carriage-corners." These cloth-covered frames are handy for many purposes, particularly for protecting and supplying some warmth to seed-pans and young seedlings.

It is essential that good drainage be given all indoor seed-pots or seed-beds. A layer of broken pots or other coarse material is placed on the bottom. Many growers place a thin layer of fine dead sphagnum moss or of peat over this drainage material, and it is useful in preventing the toc rapid drying out of the bottom of the pots. It is particularly useful in isolated pots or small boxes. Over the moss, coarse siftings from the soil may be placed, while on top only the finest and best soil should be used. The smaller the seeds, the more care must be exercised in the sowing.

The proper depth for sowing varies directly with the size of the seed. The chief advantage of very fine soil for small seeds is the greater exactness of depth of covering which it allows. Very small seeds should be sown upon the surface, which has previously been well firmed and leveled, and then covered with a very thin layer of finely sifted soil or a little old and dead moss rubbed through a sieve. This covering should be scarcely deeper than the thickness of the seeds; that is, the seeds should be barely covered. Many prefer pressing the seeds into the soil with a block. Or if one has a close propagating-box, the seeds may remain upon the surface and sufficient moisture will be supplied from the atmosphere. Such fine seeds are rarely watered directly, as even the most careful treatment would be likely to dislodge them. The soil is usually well watered before the seeds are sown, or moisture may be supplied by inserting the pot in water nearly to its rim for a few minutes. If water is applied from a rose, a thin cloth should first be spread on the soil to hold it. Celery seeds, in outdoor beds, are often sown upon a smoothly prepared surface and are then pressed in by means of the feet or a board. Some cover to prevent evaporation should be given all small seeds. This may be a board or a slate slab at first, but as soon as the plants appear glass should be substituted to admit light. (See pp. 3 to 7.)

Large seeds demand much less care as to depth of covering, as a rule. One-fourth or one-half inch is a good depth for most coarse seeds indoors. If one wishes to gauge the depth accurately, the drills may be made by a planting stick, like that shown in Fig. 16. Its flange is made of the required thickness, and it is pressed into the soil until the cap strikes the surface. This is a useful implement in seed-testing. Another device for regulating the depth of sowing, particularly in seed-testing, is the Tracy planter, shown in Fig. 17 It consists of two strips of heavy tin plate about three inches wide, hung upon two wire pivots or hinges some two inches long. At their upper edges, and equidistant from either end, the plates are joined by a firm spiral spring,

16. Planting stick.

which serves to throw the upper edges apart, and to cause the lower edges to join. The trough is now filled with the required number of seeds, and is then inserted into the earth to a given depth, when the fingers push inward on the spring and the trough opens and delivers the seeds.

Delicate seeds, which are sown out of doors, should be given a very accessible location, because they will need constant watching in dry weather and during heavy rains.

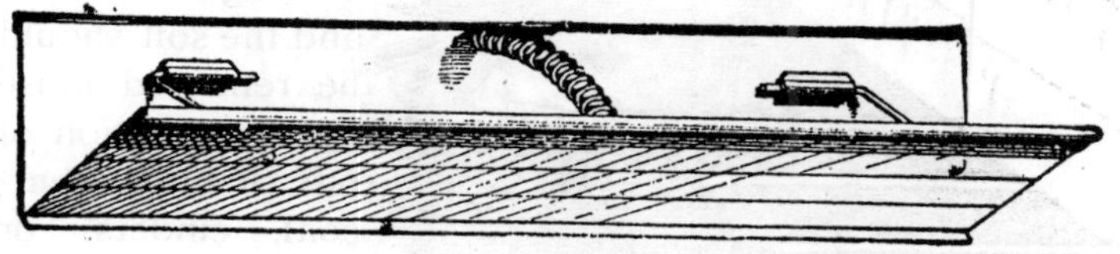

17. Tracy seed-planter.

A border along a wall is a favorite site for a seed bed. A French method of preparing such a bed is shown in Fig. 18 (after Mottet).

Damping-Off.—The gardener must always be on the lookout for the rotting-off of seedlings. This damping-off is a common ailment of young seedlings and cuttings. The stem becomes brown and constricted at or near the surface of the soil, and it soon rots and falls over. The top of the plant often remains alive and fresh for several days after it has fallen. Various fungi are concerned in this disorder, and these have recently been discussed by Atkinson (Bulletin 94, Cornell Experiment Station). The conditions which seem to particularly favor the development of these fungi are a moist and close atmosphere, crowding, and careless watering. Plants are particularly liable to damp-off if only sufficient water is applied to keep the surface moist while the under soil remains dry. Hot sand, sifted over the plants, will check it, but there is no complete remedy. As soon as the trouble appears, give more air and prick out the plants.

Spores.—Ferns, lycopodiums and selaginellas are often grown from spores. The general conditions adapted to the germination of seeds are also suitable for the germination of spores, but extra care must be taken with the drainage. If a pot is used, it should be half or more filled with drainage material, and the soil should be rendered loose by the addition of bits of brick, charcoal, cinders, or other porous materials. The surface soil should be fine and uniform. Some place a thin layer of brick dust upon the surface, in which the spores are sown. It is a frequent practice to bake the soil to destroy other spores which might cause troublesome growths. The spores should be sprinkled upon the surface and should not be covered. The pot should be set in a saucer of water, or in damp moss, and it should be covered by paper or a pane of glass if the sun strikes it. Better results are obtained if the pot or pan is placed inside a propagating-frame or under a bell-glass. In place of earth, a block or small cubes of firm peat or sandstone may be employed. The block is placed in a saucer of water and the spores are sown upon its surface. Water should not be applied directly to the spores, as it is apt to dislodge them.

18. Seed-border.

The period of germination varies in different species, but three to six weeks may be considered the ordinary limits.

While still very small, the plantlets should be pricked out, and for some time thereafter they should be subjected to the same conditions as before Spores are so exceedingly small and light that the greatest care must be exercised in growing them. In order to gather them, the fronds may be cut as soon as the sori or fruit-dots turn brown, and stored in close boxes or paper bags. When the spores begin to discharge freely, the frond may be shaken over the pot, or it may be broken up and pieces of it laid on the soil.

NOTE.—For tables of weights and longevities of seeds and quantities required for given areas, consult The Horticulturist's Rule-Book.

CHAPTER II.

SEPARATION AND DIVISION.

1. SEPARATION.

SEPARATION, or the multiplication of plants by means of naturally detachable vegetative organs, is effected by means of bulbs, bulbels, bulb-scales, bulblets, corms, tubers, and sometimes by buds.

Bulbs of all kinds are specialized buds. They are made up of a short and rudimentary axis closely encased in transformed and thickened leaves or bulb-scales. These thickened parts are stored with nutriment which is used during subsequent growth. Bulbs occur only in plants which are accustomed to a long period of inactivity. Many bulbous plants are peculiar to dry and arid regions, where growth is impossible during long intervals. A bulb is, therefore, a more or less permanent and compact leaf-bud, usually occupying the base of the stem under ground and emitting roots from its lower portion. Bulbs are conveniently divided into two great classes —the scaly, or those composed of narrow and mostly loose scales, as in the lily, and laminate or tunicate, or those composed of more or less continuous and close-fitting layers or plates, as in the onion.

19. Bulb of Lilium candidum (x⅓).

Bulbs often break up or divide themselves into two or

more nearly equal portions, as in *Lilium candidum*, shown one-third natural size in Fig. 19. The parts may be separated and treated as complete bulbs for purposes of propagation. This division or separation of bulbs proceeds in a different manner in nearly every species, yet it is so obvious that the novice need not be perplexed by it. Almost any breaking apart of these loose bulbs, if only a "heart" or central axis remains in each portion, is successful for purposes of slow multiplication; but when flowers are desired it is usually advisable to keep the bulbs as strong and compact as possible.

Bulbous plants multiply most easily by means of *bulbels*—often also called bulbules and offsets—or small bulbs which are borne about a large or mother bulb. In some lilies, as *Lilium candidum*, the bulbels form at the top or crown of the mother bulb, and a circle of roots will be found between them and the bulb; in others, as *L. speciosum* and *L. auratum*, they form on the lower part of the flower stalk. In some species the bulbels are few and very large, or even single, and they bloom the following year. In such cases the bulb undergoes a progressive movement from year to year after the manner of rootstocks, the bulb of one year bearing a more or less distinct one above and beyond it, which continues the species, while the old one becomes weak or dies. This method of bulb formation is seen in the cut of *Lilium pardalinum*, Fig. 20. In the hyacinth the bulbels form at the base of the bulb.

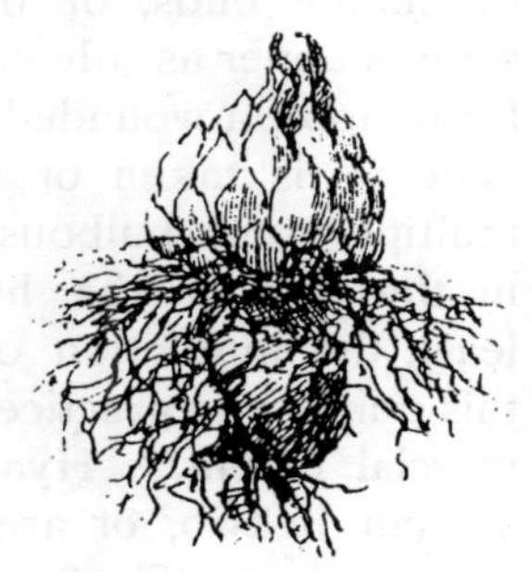

20. Bulb of Lilium pardalinum (x⅓).

Bulbels vary greatly in size and frequency in different species. Sometimes they are no larger than a grain of wheat the first year, and in other plants they are as large as hickory-nuts. In some species they are borne habitually underneath the scales of the mother bulb. These bulbels are often removed when the mother bulbs are taken up, and they are usually planted in essentially the same manner as the bulbs themselves, although it is desirable to place them, at least for the first year, in a bed or border where they may receive careful attention. Or, if they are especially small and delicate, they may be planted in pots or flats and be treated about the same as single-eye cuttings. In some lilies, the bulbels are allowed to remain attached, and the whole mass is planted in the fall in close drills. Sometimes the larger lily bulbels may produce flowers the following (or first) season, but they usually require the whole of the second season in which to complete their growth. The second fall they are ready to be permanently planted. Bulbels of some plants require even a longer time in which to mature into bulbs.

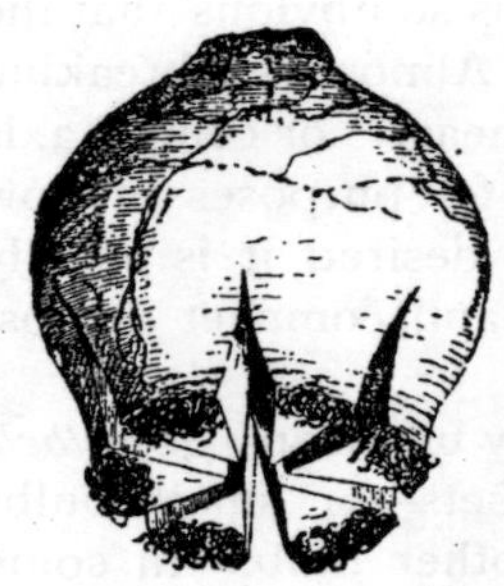

21. Cut hyacinth bulb (x½)

Bulbels are often produced by an injury to the bulb. Growth of stem and leaves is more or less checked and the energy is directed to the formation of minute buds, or bulbs, in the same manner as adventitious buds form upon a wounded stem. Advantage is taken of this fact to multiply some bulbous plants, and in the case of the hyacinths, at least, the mutilation of bulbs for this purpose is practiced to a commercial extent. Hyacinth bulbs are cut in two, or are slashed in various ways. The favorite method is to make two or three deep transverse cuts into the base of the bulb (Fig. 21). The strongest bulbs should be chosen, and the operation is performed in spring or early summer, when the bulb is taken up. The bulbs are some-

22. Hollowed hyacinth bulb (x½).

times hollowed out from the under side for half or more of their depth. This operation is sometimes performed later in the season than the other, and precaution should be exercised that the bulbs do not become too moist, else they will rot. Hollowed bulbs should be well dried before being planted. Both methods of preparing hyacinth bulbs are shown in Figs. 21 and 22, which are adapted from the Gardener's Chronicle. Fig. 23 shows a portion of the base of a cross-cut bulb, with the adventitious bulbels. The mutilated bulbs are stored during summer, and are planted in fall or spring. The wounded bulbs produce very little foliage, but at the end of the first season the bulbels will have formed. The bulbels are then separated and planted by themselves in prepared beds. Several years are required for the bulbels to mature into flowering bulbs. Some of the strongest ones may produce flowering bulbs in three years, but some of them, especially those obtained from the hollowed bulbs, will not mature short of six years. This method of propagating hyacinths is confined almost entirely to Holland.

23 Cross-cut bulb (x½).

The scales of bulbs are often employed to multiply scarce varieties. From ten to thirty of the thicker scales may be removed from the outside of the bulb without serious injury to it. These are treated in the same manner as single-eye cuttings. They are usually handled in flats or propagating-frames, and are pressed perpendicularly into a light and loose soil—half sharp sand and half leaf mold—for nearly or quite their entire length, or they may be scattered in damp moss. Keep the soil simply moist, and for hardy and half-hardy species keep the temperature rather low—from 45° to 60°. Slight bottom heat may sometimes be given to advantage. In from three to ten weeks

24. Bulb scale (x1).

a little bulbel, or sometimes two or more, will appear at the base of the scale, as shown in Fig. 24. Late autumn or early winter is a proper time for this operation. These pots or flats may be plunged outdoors during summer if the planting was done in winter, or the scales may be potted off or transferred to the open border as soon as rootlets have formed. It is the common practice with most hardy species to allow the scales to remain in the original flats during summer and to cover them the next fall, allowing them to remain outdoors over winter. The succeeding spring they are shifted into a bed or border, and by the next fall—having had two summers' growth—most species will be ready for permanent planting in the flower border.

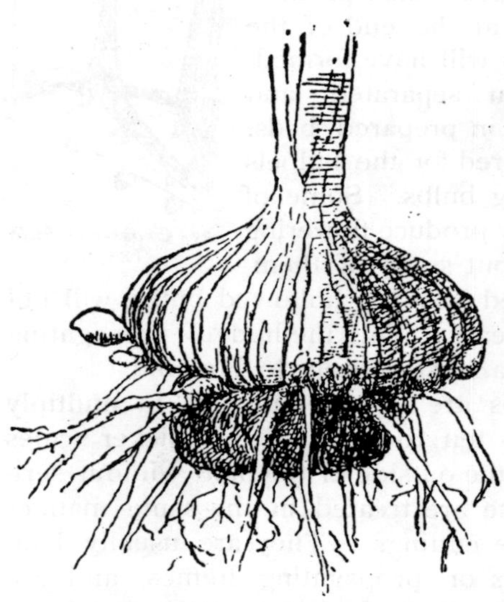

25. *Gladiolus corm* (x½).

A *bulblet* is a small bulb borne entirely above ground, usually in the axil of a leaf or in the inflorescence. Familiar examples occur in the tiger lily and in "top" onions. In the former instance, the bulblets are direct transformations of buds, while in the onion they are transformed flowers. It is impossible to draw any sharp line of separation between bulblets and buds. In some plants, certain buds detach themselves and fall to the ground to multiply the species. Sometimes these buds vegetate before they fall from the plants, as in the case of various begonias and ferns. For purposes of propagation, bulblets are treated in the same way as bulbels, and like them, they reproduce the variety upon which they grow. They will develop into full-grown bulbs in from one to three years, according to the species.

A *corm* is a bulb-like organ which is solid throughout. Familiar examples occur in the gladiolus and crocus. Cormous plants are multiplied in essentially the same manner as bulbous species. As a rule, a new corm (or sometimes two or more) is produced each year above the old one, and this commonly bears flowers the following season. This renewal is well shown in the gladiolus, Fig. 25. The illustration shows a gladiolus bottom, half size, when taken up in November. At the base are seen the withered remains of the corm which was planted in the spring, and above it the new corm, which will furnish bloom the following season. A number of *cormels* or "spawn" have also appeared about the base of the new corm. These may be planted out in a border or bed, and will produce mature bulbs in one or two seasons. The larger ones, under good treatment, will often produce bulbs an inch in diameter the first season. Some growers keep the cormels a year and a half before planting them out (that is, until the second spring), as they are thought to vegetate more evenly under such treatment; in this case they should be placed in sand to prevent too great drying out.

Adventitious cormels may be produced by various methods of wounding the mother corm, and this practice of exciting them is often necessary, as some varieties do not produce cormels freely. Each bud on the top or side of the corm may be made to produce a separate corm by cutting a deep ring around it, so as to partly divide it. Or the corm may be directly cut into as many separate pieces as there are buds or eyes, after the manner of cutting potatoes, but these pieces are usually handled in flats, where temperature and moisture can be controlled. Almost any injury to such vigorous corms as those of the gladiolus and crocus will result in the production of cormels. if care is taken that the corms do not become so cold and wet as to cause them to rot.

2. DIVISION.

The word division is commonly applied to that phase of separation in which the parts are cut or broken into pieces, in distinction to propagation by means of parts which naturally separate at the close of the season; but no hard and fast line can be drawn between the two operations. Whilst separation is mostly concerned with bulb-like and corm-like organs, division operates mostly upon tubers and rootstocks.

A *tuber* is a prominently thickened portion of a root or stem, and it is usually subterranean. The potato, sweet potato and dahlia furnish good examples. Tuberiferous plants are multiplied by planting these tubers whole, or in many cases the tubers may be cut into small portions, as described in Chapter IV., in the descriptions of cuttings. In hardy species, the tubers may be allowed to remain in the ground during winter, but they are generally dug in the fall and stored in a dry and cold place, but where they will not freeze.

An *offset* is a crown or rosette of leaves, usually borne next the surface of the ground, and which in time detaches itself and forms an independent plant. The best examples occur in the house-leeks, plants which are more familiarly known as "hen and chickens" and "man and wife." These offsets take root readily, and in propagating there is no other care necessary than to remove and plant them.

A *crown* is a detachable portion of a rootstock bearing roots and a prominent bud. Rhizomes or rootstocks multiply individuals and extend the distribution of the species by means of a progressive movement of the crowns. The rootstock grows during summer, and at the end of the season each branch develops a strong terminal bud, which usually produces a flowering stem the following season. The rootstock gradually dies away at its old extremity, and in a few years a single individual gives rise to a considerable patch. This is well shown in the common May-apple or podophyllum.

In some species these crowns are removed in the autumn, and are planted and handled in much the same manner as bulbs. The crown or "pip" of the lily-of-the valley, shown in Fig. 26, is obtained in this manner.

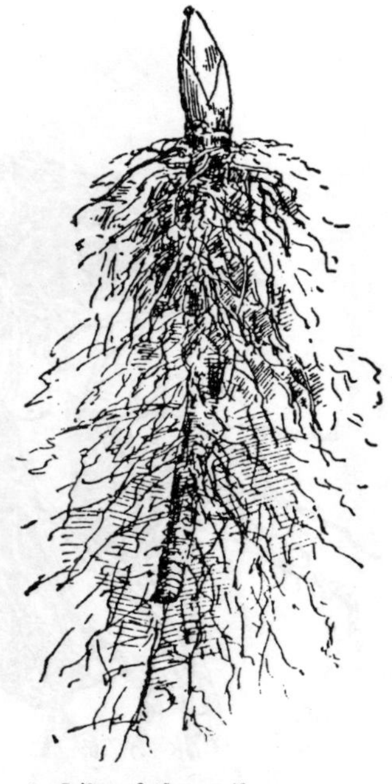

26. Lily-of-the-valley crown (x½).

Rootstocks may be divided into as many parts as there are eyes or buds, and each part is then treated as an independent plant. Familiar examples of such division are the common practices of multiplying rhubarb and canna. A canna rootstock, or "stool," is seen in Fig. 27. The observer is looking down upon the top of the stool; and the five pieces show how the operator has divided it. The two lower pieces on the left show the remains of the flower-stalks of the previous year. If the variety were very scarce, some of these pieces could be again divided into two or three.

All perennial herbs may be multiplied with more or less readiness by means of simply dividing the crowns. Most bushes may be similarly treated, as lilacs, many roses, spireas, and the like. The general stock species of herbaceous border plants — as aquilegias, hemerocallis, funkias, and the like — are generally grown in permanent small areas by nurserymen, and plants are cut out of the plot as orders are received. If, however, the nurseryman is making a special "run" on any plant, he gets his stock by dividing up the crowns or rootstocks into small portions, and then growing these for a season in specially prepared beds, or sometimes in pots.

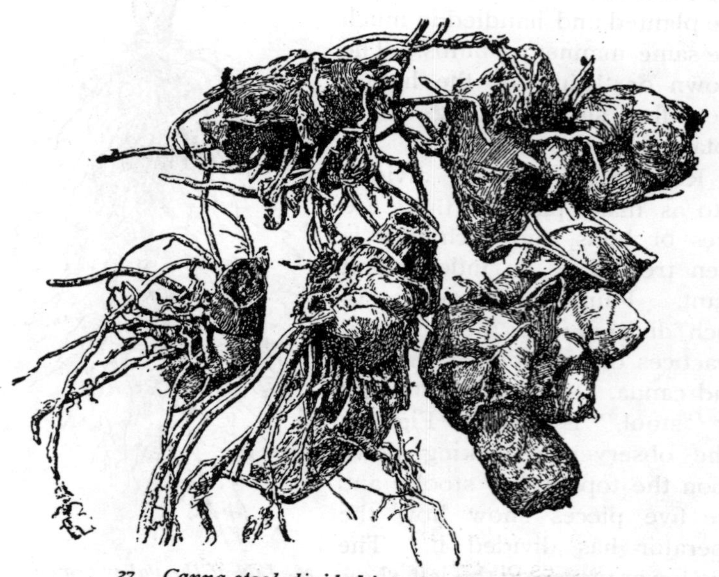

27 *Canna stool, divided into five plants* (x 1-5).

CHAPTER III.

LAYERAGE.

MANY plants habitually propagate by means of decumbent shoots and runners. These shoots become more or less covered with earth or leaves, and roots are emitted, usually at the joints. In many cases, the old shoots die away and an entirely independent plant arises from each mass of roots. In other plants, the shoots remain attached to the parent, at least for a number of years, so that the plant comprises a colony of essentially independent but connected individuals. Great numbers of plants which do not propagate naturally by means of layers are readily increased by this means under the direction of the cultivator. In most cases it is only necessary to lay down the branches, cover them with earth, and allow them to remain until roots are well formed, when the parts can be severed from the parent. Layering is one of the simplest and commonest methods of propagation, as the mother-plant nurses the layer-plants until they can sustain themselves. It is a ready means of multiplying hard-wooded plants, which do not grow well from cuttings.

All vines, and all plants which have runners or long and slender shoots which fall to the ground, may be multiplied readily by layerage. Among fruits, the black-cap raspberry and dewberry are familiar examples. The raspberry canes of the current year bend over late in summer and the tips strike the earth. If the tip is secured by a slight covering of earth, or if it finds lodgment in a mellow soil, roots are emitted, and in the fall a strong bud or "crown" or "eye" is formed for next year's growth. The parent cane is severed in the fall or spring, some 4 or 6 inches above the ground, and an independent plant, known as a "root-tip," as shown in Fig. 28, is obtained. In this instance, as in most others, it is immaterial at what point the parent stem is severed, except that a short portion of it serves as a handle in carrying the plant, and also marks the position of the plant when it is set. The black raspberry propagates itself naturally by means of these layers, and it is only necessary, in most cases, to bring the soil into a mellow condition when the tips begin to touch the ground, in order that they may find anchorage. This layering by inserting the growing point has the advantage of producing very strong "crowns" or plants in autumn from shoots or canes of the same year, and it should be more generally practiced. Even currants, gooseberries and many other plants can be handled in this way.

28. Raspberry tip (x 1-5).

In most cases of layerage, it is necessary to bend down the branches and to cover them. The covering may be

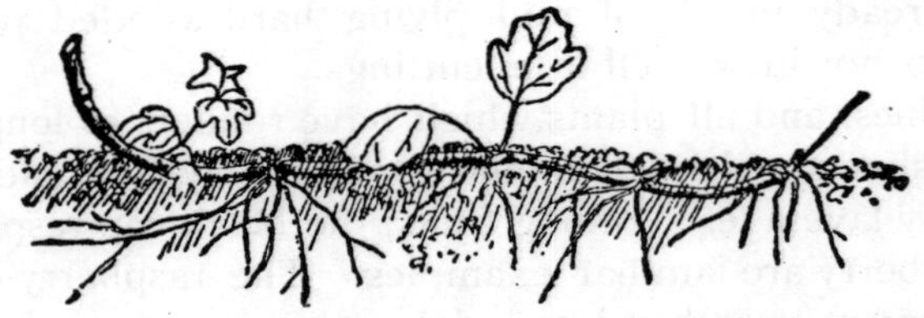

29. Covered layer of viburnum (x 1-6).

continuous, as in Fig. 29, or it may be applied only to the joints or restricted portions of the shoot, as illustrated in Fig. 30. In either case the covering should be shallow, not exceeding 2 to 5 inches. If the shoot is stiff, a stone or sod may be placed upon it to hold it down; or a crotched stick may be thrust down over it, as in the "pegging down" of propagators.

The strongest plants are usually obtained by securing only one plant from each shoot, and for this purpose the earth should be applied only at one point, preferably over a bud somewhere near the middle of the shoot. If the buds are close together, all but the strongest one may be cut out.

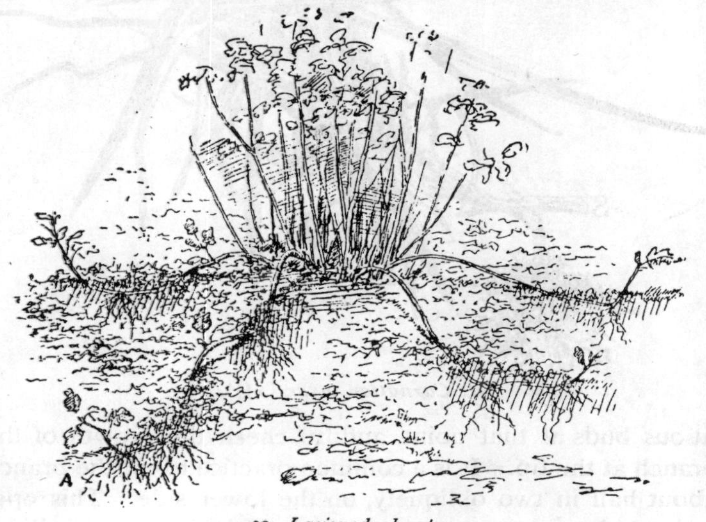

30. Layered shoots.

If more plants are desired, however, *serpentine layering* may be practiced, as shown at A in Fig. 30. The shoot is bent in an undulating fashion, and from every covered portion roots will form and a plant may be obtained. The continuously covered layer also possesses the advantage of giving more than one plant, but the roots are apt to form so continuously that definite and strong plants are rarely obtained; these rooted portions may be severed and treated as cuttings, however, with good results. The grape is sometimes propagated by serpentine layering.

Stiff and hard-wooded plants do not often "strike" or

root readily, and in order to facilitate rooting, the branch is wounded at the point where it is desired that roots shall form. This wounding serves to induce formation of adven-

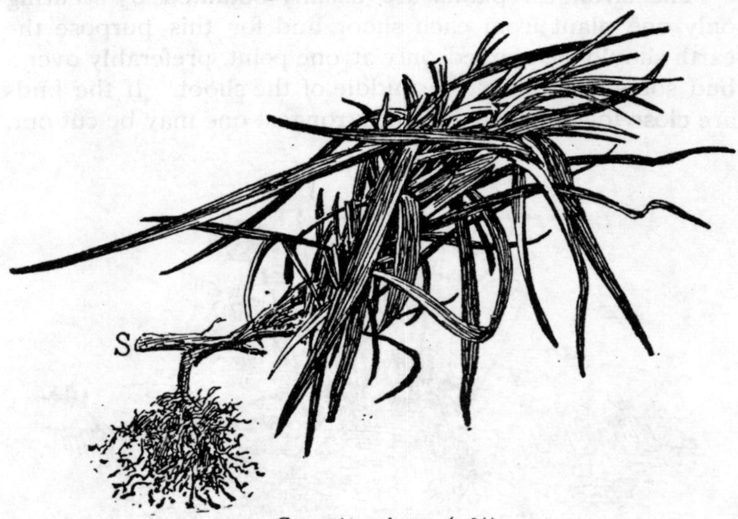

31. Carnation layer (x½).

titious buds at that point, and to check the growth of the branch at the tip. It is a common practice to cut the branch about half in two obliquely, on the lower side. This operation is known as "tongueing." "Ringing" or girdling, twisting, notching, and various other methods are employed, none of which, perhaps, possess any peculiar advantages in general practice. Some propagators cut all the buds from the covered portion. In this case the free and protruding end of the layer is expected to form the top of the new plant. "Arching," or very abrupt bending, as in serpentine layering, serves the same purpose and is the only attention necessary in most vines. A "tongued" carnation layer is shown in Fig. 31. The layered stem is at S, and the root is seen to have formed from the tongue. This method of propagating carnations is common in Europe, but the plant is always grown from cuttings in America.

When large numbers of plants are desired, as in commercial nurseries, it is often necessary to cut back the parent plant to the ground, or very nearly so, for the purpose of securing many shoots fit for layering. A plant which is cut back in the spring will produce shoots fit for layering the following spring; or some species will produce them in abundance the same year if layers of green or immature wood are desired. These parent or stock plants are called "stools" by nurserymen.

32. Mound-layering of gooseberry.

In many species, layerage is performed to best advantage by heaping earth over the stool and around the shoots. This is known as *mound* or *stool-layering*. The shoots send out roots near the base, and straight, stocky plants are obtained. The English gooseberries are almost exclusively propagated in this manner in this country. Fig. 32 shows a row of mound-layered gooseberries. The shoots are allowed to remain in layerage two years, in the case of English gooseberries, if the best plants are wanted, but in many species the operation is completed in a single season. Quinces and Paradise apple stocks are extensively mound-layered The practice is most useful in those low plants which produce short and rather stiff shoots. Sometimes these layers are severed at the end of the first season, and the plants are grown in the nursery row for a year before they are placed upon the market.

33. Layering pot.

As a rule, the best season for making layers is in spring. Rooting progresses rapidly at that season. Many plants "bleed," if layered very early in the season. Hardy shrubs may be layered in the fall, either early or late, and if an incision is made, a callus will have formed by spring. If rapid multiplication is desired, the soft and growing shoots may be layered during the summer. This operation is variously known as "summer," "herbaceous," "green" and "soft" layering. Comparatively feeble plants usually result from this practice, and it is not in common favor.

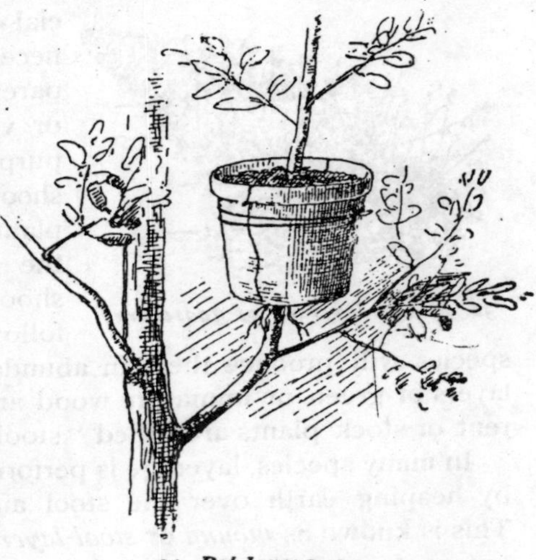

34. Pot-layerage.

In glass houses, shoots are sometimes layered in pots instead of in the earth; and the same is often done with strawberries in the field, giving the "pot-grown plants" of the nurserymen. The French have "layering-pots," with a slot in the side (Fig. 33) for the insertion of the shoot. In one style of pot, the slot extends from the rim down the entire length of the side and half-way across the bottom (Fig. 36.)

Pot-layering, *circumposition*, *air-layering* and *Chinese layering* are terms applied to the rooting of rigid stems by means of surrounding them, while in their natural position, with earth or moss, or similar material. The stem is wounded—commonly girdled—and a divided pot or box is placed about it and filled with earth (Fig. 34). The roots start from above the girdle, and when they have filled the pot the stem is severed, headed back, and planted. Pot-layering is practiced almost exclusively in greenhouses, where it is possible to keep the earth uniformly moist. But even there it is advisable to wrap the pot in moss to check evaporation from the soil. Some plants, like *Ficus elastica*, can be readily rooted by wrapping them with moss alone, if the atmosphere is sufficiently close. A paper cone may be used in place of a pot where the atmosphere is not too humid, as in carnation houses (Fig. 35). Pot-layering is employed not only for the purpose of multiplying plants, but in order to lower the heads of "leggy" or scraggly specimens. The pot is inserted at the required point upon the main stem, and after roots have formed abundantly the top may be cut off and potted independently, the old stump being discarded.

35. Air-layering in a paper cone (x½).

36. Layering-pot.

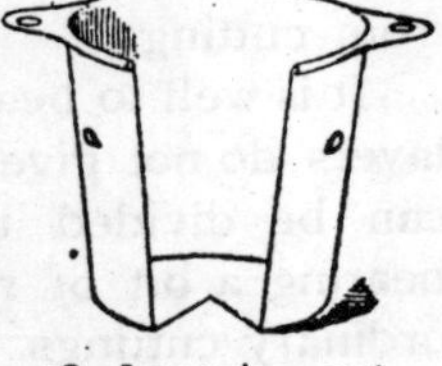

38. Layering-cup.

The French have various handy devices for facilitating pot-layering. Fig. 36 shows a layering-pot, provided with a niche in the side to receive the stem, and a flange behind for securing it to a support. The pot shown in Fig. 33 is a similar device. Fig. 37 represents a layering-cone. It is made of zinc or other metal, usually 4 or 5 inches high, and is composed of two semi-conical wings, which are hinged on the back and are secured in front, when the instrument is closed, by means of a hinge-pin. A cord is inserted in

37 Layering-cone.

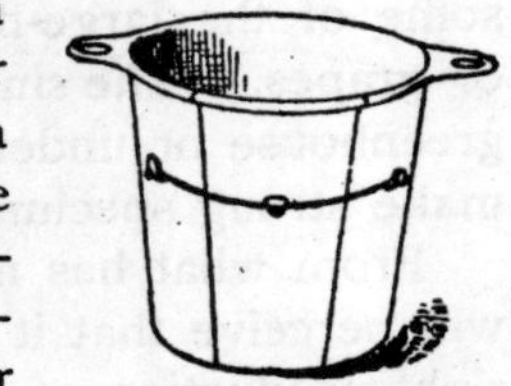

39. Layering-cup.

one side, with which to hang it on a support. A cup or pot with a removable side is also used. This is shown open in Fig. 38 and closed in Fig. 39. An ingenious compound layering-pot is shown in Fig. 40. The main stem or trunk of the plant is carried through the large opening, and the branches are taken through the smaller pots at the side. Kier's layering-boxes or racks are shown in Figs. 41 and 42. The trays are filled with earth or moss, and the branches are laid in through the chinks in the border and are treated in the same manner as ordinary outdoor layers. These racks supply a neat and convenient means of increasing greenhouse plants which do not readily strike from cuttings.

40. Compound layering-pot

It is well to bear in mind that when layers do not give strong plants, they can be divided into portions, each bearing a bit of root, and treated as ordinary cuttings. This is an important operation in the case of rare varieties which are multiplied by means of soft or green layers, as some of the large-flowered clematises and new varieties of grapes. The small, weak plants are handled in a cool greenhouse or under frames, usually in pots, and they soon make strong specimens.

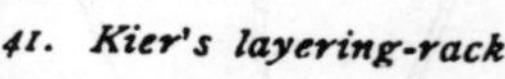

41. Kier's layering-rack

From what has now been said of layerage, the reader will perceive that it may be employed either for the outright production of new plants, or as a means of starting or "striking" plants. In the latter case, the layer plants, after having been separated from the parent plant, are set in nursery rows and there grown for one season; and in this way stronger and more shapely plants may be obtained. As a general statement, it may be said that all bush-like or vine-like plants which do not strike readily from cuttings, nor produce seeds freely, or of which the seeds are very slow to germinate, are usually multiplied by layerage.

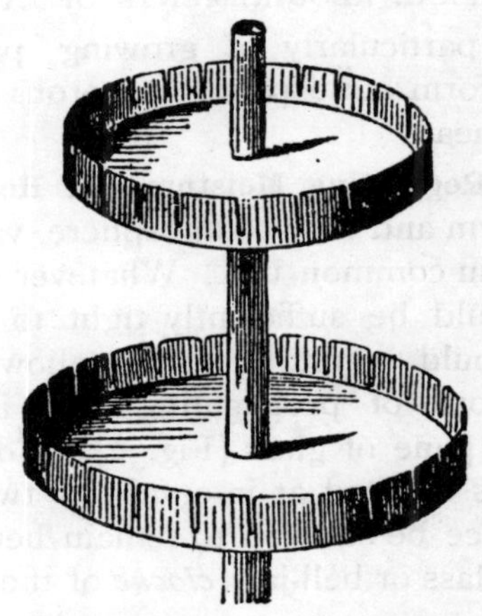

42. Kier's circular layering-rack

CHAPTER IV.

CUTTAGE.

1. GENERAL REQUIREMENTS OF CUTTINGS.

CUTTINGS, particularly of growing parts, demand a moist and uniform atmosphere, a porous soil, and sometimes bottom heat.

Devices for Regulating Moisture and Heat.—In order to secure a uniform and moist atmosphere, various propagating-frames are in common use. Whatever its construction, the frame should be sufficiently tight to confine the air closely; it should admit light, and allow of ventilation. The simplest form of propagating-frame is a pot or box covered with a pane of glass (Fig. 2). To admit of ventilation, the glass is tilted at intervals, or two panes may be used and a space be allowed to remain between them. A common bell-glass or bell-jar (*cloche* of the French) makes one of the best and handiest propagating-frames, because it admits light upon all sides and is convenient to handle. It is particularly serviceable in the propagation of tropical or "stove" plants; and it is in universal use for all difficult and rare subjects which are not propagated in large numbers. A hand-glass or hand-light (Fig. 44)

44 Hand-glass.

answers the same purpose and accommodates a larger number of plants. A useful propagating-box for the window garden or amateur conservatory is shown in Fig. 45. A box 2 or 3 inches high is secured, and inside this a zinc or galvanized iron tray is set, leaving sufficient space between it and the box to admit a pane of glass upon every side. These panes form the four sides of the box, and one or two panes are laid across the top. The metal tray holds the soil and allows no water to drip upon the floor. One of the best boxes for general purposes is made in the form of a simple board box without top or bottom, and 15 or 18 inches high, the top being covered with two sashes, one of which raises upon a hinge (Fig. 46). Four by three feet is a convenient size. An ordinary light hotbed frame is sometimes constructed upon the bench of a greenhouse and covered with common hotbed sash. Propagating-houses are sometimes built with permanent propagating-frames of this character throughout their length, as shown in Fig. 47. Such permanent frames are mostly used for conifers, either from cuttings or grafts (usually the latter) and also for grafts of rhododendrons (See Chapters V. and VI.).

45. Small propagating-box.

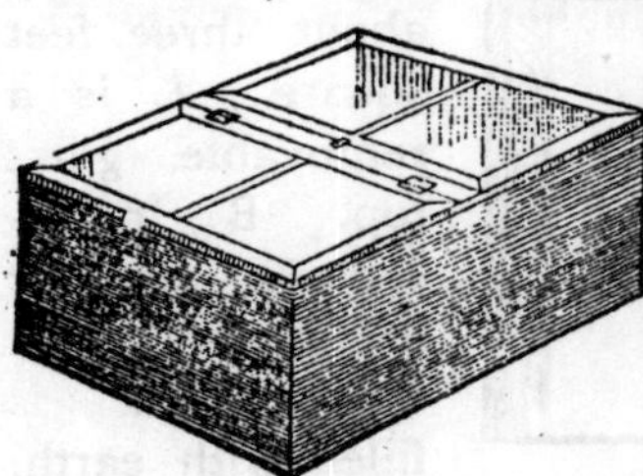

46. Propagating-box.

In all the above appliances heat is obtained from the sun or from the bench-pipes or flues of a greenhouse. There are various contrivances in which the heat is applied locally, for the purpose of securing greater or more uniform heat. One of the simplest and best of these is the propagating-oven shown in Fig. 48. It is a glass-

47. Permanent propagating-frames in a greenhouse.

covered box about two feet deep, with a tray of water beneath the soil, and which is heated by a lamp. A similar but somewhat complicated apparatus is illustrated in Figs. 49, 50, 51. This is an old form of oven, which has been variously modified by different operators. Fig. 49 shows a sectional view of the complete apparatus. The box, A A, is made of wood, and is usually about three feet square. L is a removable glass top. B represents a zinc or galvanized iron tray which is filled with earth, in which seeds are sown or pots are plunged. C is a water tray, to which the water is applied by means of a funnel extending

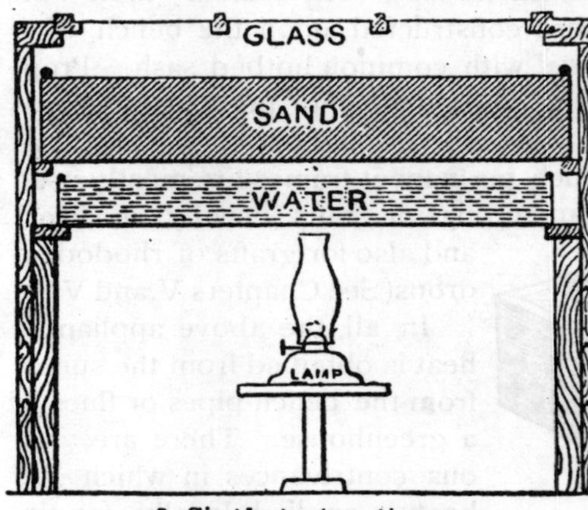

48. Simple propagating-oven.

through the box. A lamp, D, supplies the heat. A funnel of tin, *e e*, distributes the heat evenly. Holes should be provided about the bottom of the box to admit air to the flame. A modified form of this device is shown in Figs. 50 and 51. The water tray, G, slides in upon ledges, so that it can be removed, and the heat funnel, L D L, slides in similarly and is made to surround the flame like a chimney.

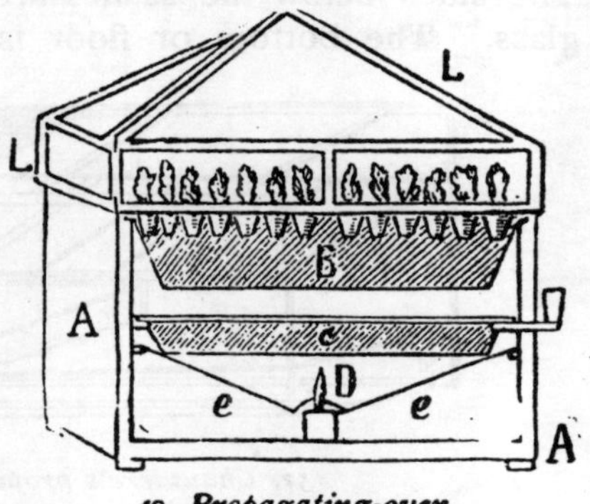

49. Propagating-oven.

The front side of the apparatus is removable, and the top of the frame, K, is made of metal. The cover for this apparatus is shown in Fig. 51. The ends, *a a*, are made of wood, with openings, indicated by the arrows, to allow of ventilation. The front and top, *g g*, are made of glass. The frame-work, *c c c*, is made of metal. The cover is hinged on, or held with pegs, I I, Fig. 50.

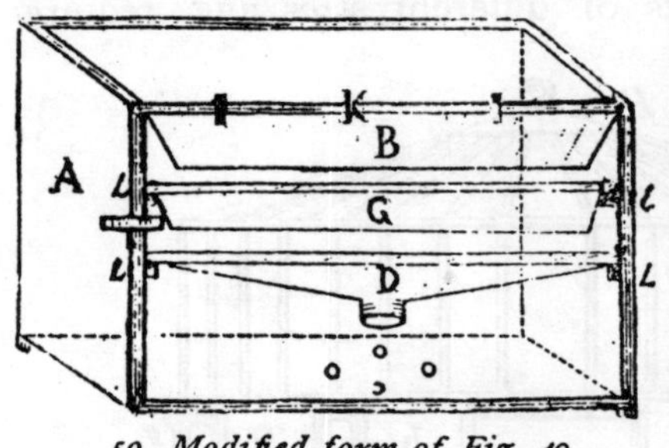

50. Modified form of Fig. 49.

Chauvière's propagating-oven, a French apparatus, is shown in Fig. 52. It is essentially a miniature greenhouse. The sashes are seen at *c c*, and over them is a cloth or matting screen (shown at the right).

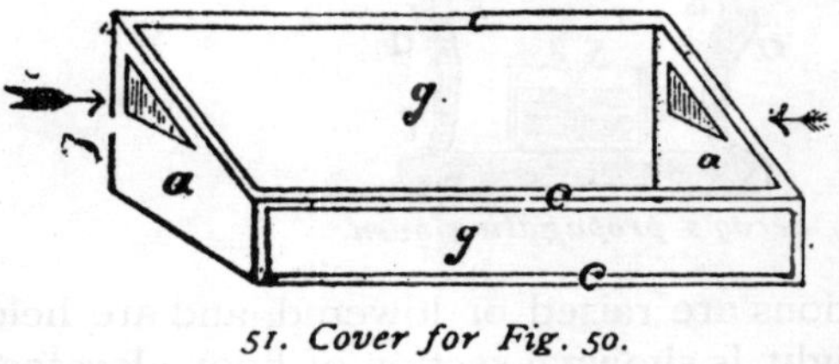

51. Cover for Fig. 50.

The sides below the sashes are enclosed, preferably with glass. The bottom or floor is movable, and it is some-

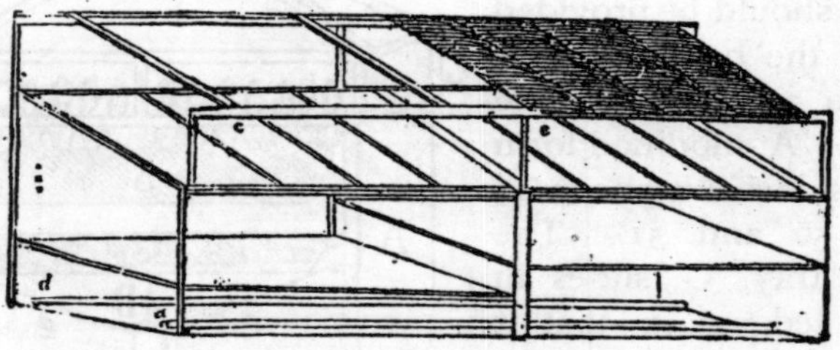

52. Chauvière's propagating-oven.

times divided into two or three sections, to allow of the accommodation of plants of different sizes and require-

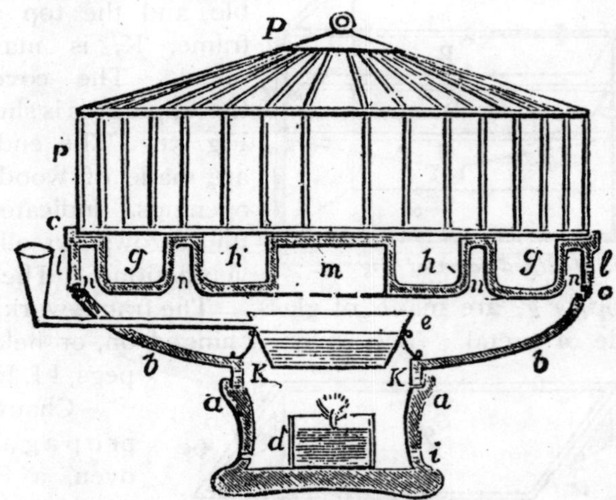

53. Lecoq's propagating-oven.

ments. These sections are raised or lowered, and are held by pegs. At the right is shown a section of floor elevated, and at the left another section occupying a lower position. Heat is supplied usually by hot water in the tubes, *d d.* A very elaborate circular French device, known as Lecoq's propagating-oven, is illustrated in Fig. 53. It is an interesting apparatus, and is worth attention as showing the care which has been taken to control the conditions of vegetation and germination. It is too elaborate for common purposes, and yet for the growing of certain rare or difficult subjects it might find favor among those who like to experiment; and it affords an accurate means of studying plant growth under control. The apparatus is sold in France for about $6. All the portion below the glass top, *P*, *P*, is made of earthenware. The base, *a a*, holds a lamp, *d*; *e* is a water reservoir, to which water is supplied by means of the funnel, *j*. A vase or rim, *b b*, rests upon the base, and upon it a plate or disc, *c c*, is fitted. Above this is the glass top, *P P*. Air is admitted to the apparatus at *i*, *K K*, and between the vase and plate, as at *c* on the right. The plate contains two concentric circular grooves, *g g* and *h h*. In these grooves the soil is placed or pots plunged. The heat circulates in the valleys *m* and *n n n n*, and supplies a uniform temperature to both sides of the plants.

Barnard's propagating-tank, Fig. 54, is a practicable device for attachment to a common stove. A similar apparatus may be attached to the pipes of a greenhouse. The tank consists of a long wooden box made of matched boards, and put together with paint between the joints to make it water-tight. The box should be about 3 feet wide and 10 inches deep, and may be from 10 to 30 feet long, according to the space required. In the middle of the box is a partition, extending nearly the whole length, and on the inside, on each side, is a ledge or piece of moulding to support slate slabs to be laid over the entire surface of the box. The slates are supported by the ledges and by the central partition, and should be fastened down with cement to prevent the propagating-sand from falling into the tank. One slate is left out near the end, next the fire, to enable the operator to see the water and to keep it at the right level. On the slates sand is spread, in which the cuttings may be struck, the sand nearly filling the box. At one end of the box is placed a common cylinder stove, with smoke-pipe to the chimney. Inside the stove is an iron pipe, bent in a spiral. This coil, which is directly in the fire, is connected by

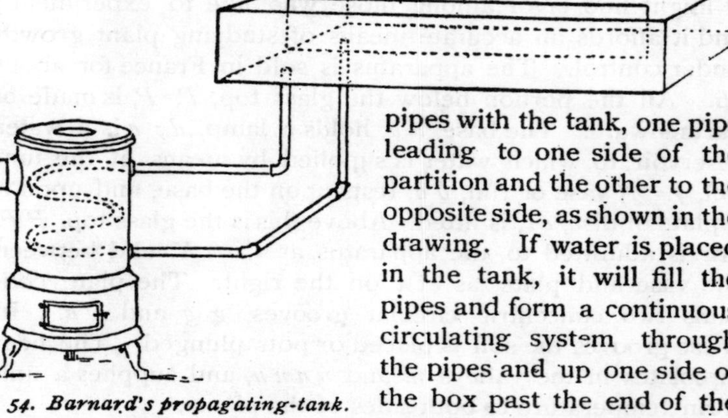

54. *Barnard's propagating-tank.*

pipes with the tank, one pipe leading to one side of the partition and the other to the opposite side, as shown in the drawing. If water is placed in the tank, it will fill the pipes and form a continuous circulating system through the pipes and up one side of the box past the end of the partition, and down the other side. A fire in the stove causes the water to circulate through the tank and impart to the bed a genial warmth.

There are various tanks designed to rest upon the pipes in a greenhouse. The principle of their construction is essentially the same as of those described in previous pages, —bottom heat, a tray of water, and a bed of soil. Earthenware tanks are commonly employed, but a recent English device, Fig. 55, is made of zinc. It is about 7 inches deep, and holds an inch or two of water in the bottom. A tray 5 inches deep sets into the tank. The water is supplied through a funnel at the base.

Cuttings usually "strike" better when they touch the side of the pot than when they are wholly surrounded by soil. This is probably because the earthenware insures greater uniformity in drainage than the earth, and supplies air and a mild bottom heat; and it is possible that the deflection of the plant ſood towards the side of the pot, because of evaporation therefrom, induces better growth at that point. Various devices are employed for the purpose of securing these advantages to the best effect. These are usually double pots, in one of which water is placed. A good method is that represented in Figure 56, which shows a pot, *b*, plugged with plaster of

55 *Zinc propagating-tank.*

Paris at the bottom, placed inside a larger one. The earth is placed between the two, drainage material occupying the bottom, *a*, and fine soil the top, *c*. Water stands in the inner pot as high as the dotted line, and feeds uniformly into the surrounding soil. The positions of the water and soil are frequently reversed, but in that case there is less space available for cuttings. A double pot, with moisture supplied in a surrounding cushion of sphagnum moss, is seen in Fig. 1. Neumann's cutting-pot is shown in

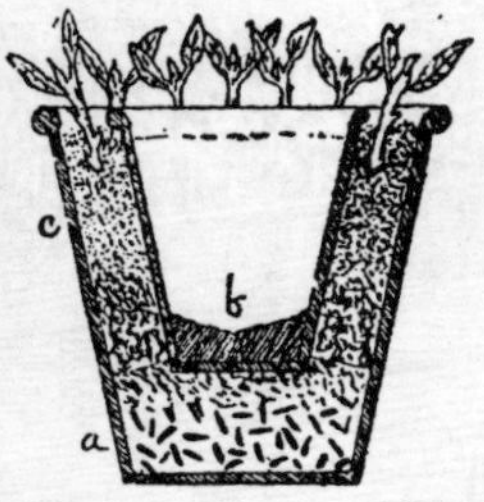

56. *Forsyth's cutting-pot.*

Fig. 57. This contains an inverted pot in the center, *a*, designed to supply drainage and to admit heat into the center of the mass of soil. A good method of striking difficult subjects is as follows: Fill a saucer with moss. Upon this place an inverted flower-pot. Insert the cutting through the hole in the bot-

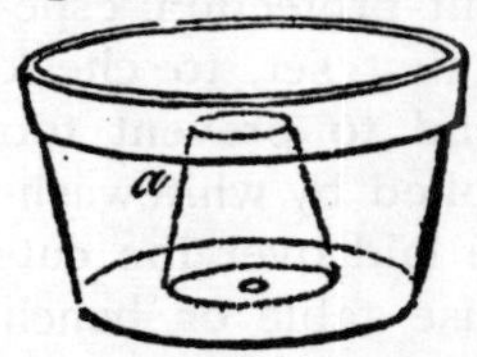

57. *Neumann's cutting-pot.*

tom of the pot, so that it stands in the moss and almost touches the saucer. Keep the moss moist.

Some kind of protection, commonly combined with bottom heat, is always given cuttings made from the soft and growing parts. In indoor work, any of the devices named above may be employed, but a box like that shown in Fig. 46 is one of the most useful for common operations. Or

58. Cutting-bench shaded with lath.

the greenhouse itself may afford sufficient protection, especially if the cuttings are shaded when first set, to check evaporation from the plant and soil, and to prevent too great heat. This shading is usually supplied by whitewashing the glass, or a newspaper may be laid over the cutting-bed for a few days. A greenhouse table or bench prepared for the growing of cuttings is known as a "cutting-bench." If the cuttings become too dry or too hot, they will wilt or "flag." A good cutting-bench should be near the glass, and either exposed to the north or else capable of being well shaded. A good bench, facing south and shaded over the glass with a lath screen, is illustrated in Fig. 58. The details of soils are discussed on the following page.

In outdoor work, soft cuttings are usually placed in an ordinary coldframe, and these frames must be shaded. They may be placed under trees or on the shady side of a building, or if they are numerous, as in commercial establishments, a cloth screen should be provided, as shown in Fig. 7, page 6.

Bottom Heat is always essential to the best success with cuttings. In outdoor work, this is supplied by the natural heat of the soil in spring and summer, and it is often intensified by burying hard-wooded cuttings bottom end up for a time before planting them. This operation of inverting cuttings is often practiced with grapes, particularly with the Delaware and others which root with some difficulty. The cuttings are tied in bundles and buried in a sandy place, with the tops down, the butts being covered two or three inches with sand. They may be put in this position in the fall and allowed to remain until the ground begins to freeze hard, or they may be buried in spring and allowed to remain until May or June and then be regularly planted. In outdoor cuttage, the cuttings which are of medium length, from 6 to 8 inches, derive more bottom heat than the very long ones, such as were formerly used for the propagation of the grape. In indoor work, bottom heat is obtained by means of fermenting manure, or, preferably, by greenhouse pipes. Cutting-benches should have abundant piping beneath, and in the case of many tropical and sub-tropical species the bottom heat may be intensified by enclosing the benches below, so that no heat can escape into the walks. Doors can be placed in the partition alongside the walk, to serve as ventilators if the heat should become too intense. In all cuttings, bottom or root growth should precede top growth, and this is aided by bottom heat.

Soils.—Soil for all cuttings should be well drained. It should not be so compact as to hold a great quantity of water, nor should it be so loose as to dry out very quickly. It should not "bake" or form a crust on its surface. As a rule, especially for cuttings made of growing parts, the soil should not contain fresh vegetable matter, as such material holds too much water and is often directly injurious to the cutting, and it is likely to breed the fungi of damping-off. A coarse, sharp, clean sand is the best material for use indoors. Very fine sand packs too hard, and should rarely be used. Some propagators prefer to use fine gravel, composed of particles from an eighth to a fourth of an inch in diameter, and from which all fine material has been washed. This answers well for green cuttings; but a propagating-frame should be used to check evaporation, and attention be given to watering, because drainage is so perfect and the material so quickly permeable that uniformity of treatment is thereby secured. Damping-off is less liable to occur in such material than in denser soils. The same advantages are to some extent present in sphagnum moss and cocoanut fiber, both of which are sometimes used in place of earth. The "silver sand" used by florists is a very clean and white sand, which derives its particular advantages from the almost entire absence of any vegetable matter. But it is not now considered so essential to successful propagation as it was formerly, and fully as good material may often be found in a common sand-bank. Cuttings which strike strongly and vigorously may be placed in a soil made of light garden loam with twice its bulk of sand added to it. All soils used for indoor cuttage should be sifted or screened before using, to bring them to a uniform texture.

Hard-wood cuttings are commonly planted outdoors in mellow and light garden loam, well trenched. Only fine and well-rotted manure should be applied to the cutting-bed, and it should be well mixed with the soil. In most cases, a well-drained soil gives best results, but some cuttings root and grow well in wet soils, or even in standing water, as poplars, willows, some of the dogwoods, plane-tree, and others.

The Formation of Roots.—As a rule, roots arise most readily from a joint, and it is, therefore, a common practice to cut off the base of the cutting just below a bud, as shown in the grape cutting, Fig. 59. Sometimes the cutting is severed at its point of attachment to the parent branch, and a small portion, or "heel," of that branch is allowed to remain on the cutting. This heel may be nothing more than the curved and hardened base of the cutting at its point of attachment, as in the cornus cutting, Fig. 60. Sometimes an entire section of the parent branch is removed with the cutting, as in the "mallet" cuttings of grapes, Fig. 61. Of course, comparatively few heel or mallet cuttings can be made from a plant, as only one cutting is obtained from a branch, and it is advisable, therefore, to "cut to buds" rather than to "cut to heels;" yet there are many plants which demand a heel, if the most satisfactory results are to be obtained. The requirements of the different species in this regard can be learned only by experience; but it may be said that in general the hardest or closest wooded plants require a heel or a joint at the base. Willows, currants, basswoods, and others with like soft wood, emit roots readily between the buds, yet even in these cases propagators generally cut to buds.

59. Grape cutting (x ½)

Wounds upon plants begin to heal by the formation of loose, cellular matter which gives rise to a mass of tissue known as a *callus*. This tissue eventually covers the entire wound, if complete healing results. As a rule, the first apparent change in a cutting is the formation of a callus upon the lower end, and it is commonly supposed that this process must be well progressed before roots can form.

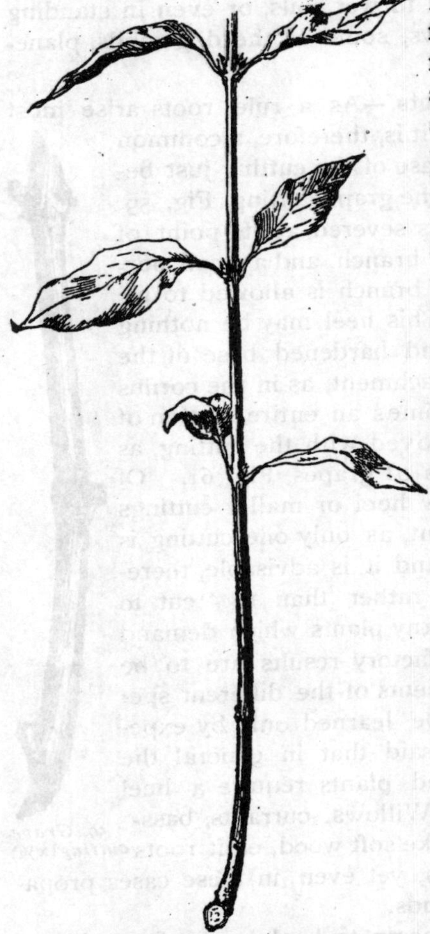
60. *Heel cutting of cornus* (x½).

But roots do not arise from the callus itself, but from the internal tissue, and in many plants they appear to bear no relation to the callus in position. In willows, for instance, roots arise from the bark at some distance from the callus. Yet, as a matter of practice, best results are obtained from callused cuttings, particularly if the cuttings are made from mature wood, but this is probably due to the fact that considerable time is required for the formation of the adventitious buds which give rise to the roots, not to any connection between the callusing and rooting processes themselves.

Hard-wood cuttings give better results when kept dormant for some time after they are cut. They are usually made in the fall, and stored during the winter in sand, sawdust or moss in a cool cellar, or buried in a sandy and well-drained place. This, at least, is the practice with hard-wood cuttings of deciduous plants, like currants, gooseberries, grapes, and many ornamental trees and shrubs. Hard-wood evergreen cuttings, when taken in the fall, are usually set at once, as their foliage will not allow them to be buried with safety; but in this case, the cuttings are kept "quiet" or dormant for a time, to allow callusing to progress. If cuttings are buried so deep that they cannot sprout, callusing may be hastened by placing them in a mild temperature. Single-eye grape cuttings are sometimes packed between layers of sand in a barrel, and the barrel is set under a forcing-house bench where the temperature is about 50°. Eight or ten inches of sand is usually placed over the top layer. In this manner, cuttings which have been obtained in winter or early spring may be callused before planting time.

It is a singular fact that the lower end of the cutting, as it stood upon the parent plant, produces roots, and the upper end produces leaves and shoots, even if the cutting is inverted. And if the cutting is divided into several parts, each part will still exhibit this same differentiation of function. This is true even of root cuttings, and of other cuttings which possess no buds. The reasons for this localization of function are not clearly understood, although the phenomenon has often been the subject of study. Upon this fact depends the hastening of the rooting process in inverted cuttings by the direct application of heat to the bottoms, and it likewise indicates that care must be taken to plant cuttings in approximately their natural direction if straight and handsome plants are desired. This remark applies particularly to horse-radish "sets," for if these are placed wrong end up (even though they are root cuttings), the resulting root will be very crooked.

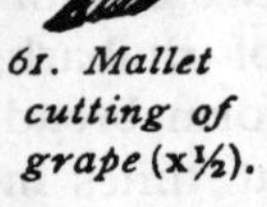

61. Mallet cutting of grape (x½).

The particular method of making the cutting, and the treatment to which it should be subjected, must be determined for each species or genus. Some plants, as many maples, can be propagated from wood two or three years old, but in most cases the wood of the previous or present season's growth is required. Nearly all soft and loose-wooded plants grow readily from hard-wood cuttings, while those with dense wood are generally multiplied more easily from soft or growing wood. Some plants, as oaks and nut-tress, are propagated from cuttings of any description only with great difficulty, although the hickories grow rather freely from soft tip-cuttings of roots. It is probable, however, that all plants can be multiplied by cuttings if properly treated. It often happens that one or two species of a closely defined genus will propagate readily from cuttings while the other species will not, so that the propagator comes to learn by experience that different treatment is profitable for very closely related plants. For instance, most of the viburnums are propagated from layers in commercial establishments, but *V. plicatum* (properly *Viburnum tomentosum*) is grown extensively from cuttings.

2. THE VARIOUS KINDS OF CUTTINGS.

Cuttings are made from all parts of the plant. In its lowest terms, cuttage is a division of the plant itself into two or more nearly equal parts, as in the division of crowns of rhubarb, dicentra, and most other plants which tend to form broad masses or stools. This species of cuttage is at times indistinguishable from separation, as in the dividing of lily bulbs (page 27), and at other times it is essentially the same as layerage, as in the dividing of stools which have arisen from suckers or layers. This breaking or cutting up of the plants into two or more large parts which are already rooted is technically known as Division, and is discussed in Chapter II. It is only necessary, in dividing plants, to see that one or more buds or shoots remain upon the portions, and these portions are then treated in the same manner as independent mature plants are, or sometimes, when the divisions are small and weak, they may be handled for a time in a frame or forcing-house in the same manner as ordinary cuttings.

Cuttings proper may be divided into four general classes, with respect to the part of the plant from which they are made: 1, of tubers; 2, of roots and rootstocks; 3, of stems; 4, of leaves. All these forms of cuttings reproduce the given variety with the same degree of certainty that grafts or buds do.

Tuber Cuttings.—Tubers are thickened portions of either roots or stems, and tuber cuttings, therefore, fall logically under those divisions; but they are so unlike ordinary cuttings in form that a separate classification is desirable. Tubers are stored with starch, which is designed to support or supply the plant in time of need. Tuber cuttings are, therefore, able to support themselves for a time if they are placed in conditions suited to their vegetation. Roots rarely arise from the tubers themselves, but from the base of the young shoots which spring from them. This fact is familiarly illustrated in the cuttings of Irish and sweet potatoes. The young sprouts can be removed and planted separately, and others will arise from the tuber to take their places. This practice is employed sometimes with new or scarce varieties of the Irish potato, and three or four crops of rooted sprouts can be obtained from one tuber. The tuber is cut in two lengthwise, and is then laid in damp moss or loose earth with the cut surface down, and as soon as the sprouts throw out roots sufficient to maintain them they are severed and potted off. Sweet potatoes are nearly always propagated in this manner.

In making tuber cuttings, at least one eye or bud is left to each piece, if eyes are present; but in root-tubers, like the sweet potato, there are no buds, and it is only necessary to leave upon each portion a piece of the epidermis, from which adventitious buds may develop. The pseudo-bulbs of some orchids are treated in this manner, or the whole bulb is sometimes planted. A shoot, usually termed an off-shoot, arises from each pseudo-bulb or each piece of it, and this is potted off as an independent plant. (See Orchids, in Chapter VI.)

Cuttings made from the ordinary stems of some tuberiferous plants will produce tubers instead of plants. This is the case with the potato. The stem cutting produces a small tuber near its lower extremity, or sometimes in the axil of a leaf above ground, and this tuber must be planted to secure a new plant. Leaf cuttings of some tuberiferous or bulbiferous plants produce little tubers or bulbs in the same way (see the gloxinia, Fig. 81). Hyacinth leaves, inserted in sand in a frame, will soon produce little bulblets at their base, and these can be removed and planted in the same manner as the bulbels described in Chapter II.

Many tubers or tuber-like portions, which possess a very moist or soft interior and a hard or close covering, vegetate more satisfactorily if allowed to dry for a time before planting. The pseudo-bulbs of orchids, crowns of pine-apples, and cuttings of cactuses are examples. Portions of cactuses and pine-apples are sometimes allowed to lie in the sun from two to four weeks before planting. This treatment dissipates the excessive moisture, and induces the formation of adventitious buds.

62. Root cutting of blackberry (x½).

Root Cuttings. — Many plants can be multiplied with ease by means of short cuttings of the roots, particularly all species which possess a natural tendency to "sucker" or send up sprouts from the root. All rootstocks or underground stems can be made into cuttings, as explained under Division, in Chapter II.; but true root cuttings possess no buds whatever, the buds developing after the cutting is planted. Roots are cut into pieces from 1 to 3 inches long, and are planted horizontally in soil or moss. These cuttings thrive best with bottom heat, but blackberries and some other plants grow fairly well with ordinary outdoor treatment. A root cutting of the blackberry is shown in Fig. 62. (See Blackberry, in Chapter VI.) A growing dracæna root cutting is exhibited in Fig. 63. The cuttings of this plant are handled in a propagating-frame or on a cutting-bench in a warm greenhouse. The bouvardias and many other plants are grown in the same manner. Many of the fruit trees, as peach, cherry, apple and pear, can be grown readily from these short root cuttings in a frame. Among kitchen garden plants, the horse-radish is the most familiar example of propagation by root cuttings. The small side roots, a fourth inch or so in diameter, are removed when the horse-radish is dug in fall or spring, and are cut into 4 to 6-inch lengths, as seen in Fig. 64. These cuttings are known as "sets" among gardeners. (See horse-radish, Chapter VI.) When the crowns of horse-radish are cut and used for propagation, the operation falls strictly under division, from the fact that buds or eyes are present; and the same remark applies to rhubarb,

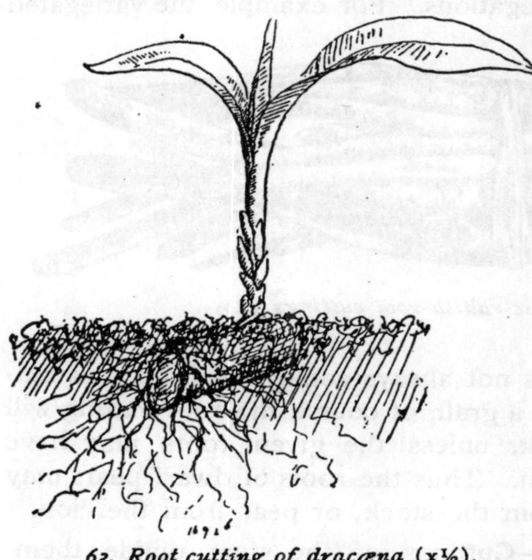

63. Root cutting of dracæna (x½).

which, however, is not propagated by true root cuttings.

Whilst root cuttings perpetuate the variety, they do not always transmit variegations. For example, the variegated

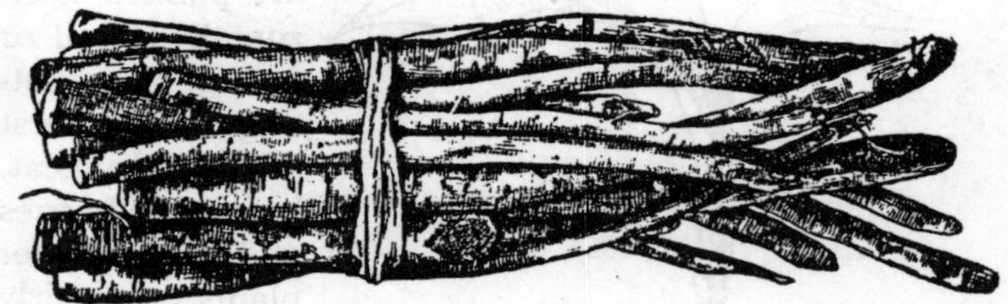

64. Horse-radish root cuttings (x½).

prickly comfrey does not always come true from root cuttings. If the top is a graft, of course the root cutting will reproduce the stock, unless the given roots may have started from the cion. Thus the roots of dwarf pears may be either quince from the stock, or pear from the cion.

Stem Cuttings.—Cuttings of the stem divide themselves into two general classes: those known as cuttings of the ripe, mature or hard wood, and cuttings of the green, immature or soft wood. The two classes run into each other, and no hard and fast line can be drawn between them.

Hard-wood cuttings are made at any time from late summer to spring. It is advisable to make them in the fall, in order to allow them to callus before the planting season, and to forestall injury which might result to the parent plant from a severe winter. They may be taken as early as August, or as soon as the wood is mature, and be stripped of leaves. Callusing can then take place in time to allow of fall planting. Or, the cuttings taken in early fall may be planted immediately, and be allowed to callus where they stand. All fall cutting-beds should be mulched, to prevent the heaving of the cuttings. As a rule, however, hard-wood cuttings are buried on a sandy knoll or are stored in moss, sand or sawdust in a cellar until spring. (See page 56.)

There is no general rule to govern the length of hard-wood cuttings. Most propagators prefer to make them 6 to 10 inches long, as this is a convenient length to handle, but the shorter length is preferable. Two buds are always to be taken, one bud or one pair at the top and one at the bottom, but in "short-jointed" plants more are obtained. Sometimes all but the top buds are removed to prevent the starting of shoots or sprouts from below the soil. Grape cuttings are now commonly cut to two or three buds (Fig. 59), two being the favorite number for most varieties. (See grape, Chapter VI.) Currant and gooseberry cuttings (Fig. 65) usually bear from 6 to 10 buds. All long hard-wood cuttings are set perpendicularly, or nearly so, and only one or two buds are allowed to stand above the surface.

When the stock is rare, cuttings are made of single eyes or buds. This is particularly the case with the grape (see Chapter VI.), and currants and many other plants are occasionally grown in the same manner. Fig. 66 shows a single-eye grape cutting. Such cuttings, whatever the species, are commonly started under glass with bottom heat, either upon a cutting-bench or in a hotbed, being planted an inch or so deep in a horizontal position, with the bud up. The soil should be kept uniformly moist, and when the leaves appear the plants should be frequently sprinkled. In from 30 to 40 days the plants are ready to pot off. Single-eye cuttings are usually started about three or four months before the season is fit for outdoor planting, or about February in the northern states. The most advisable method of treatment varies with the season and locality, as well as with the species or

65. Currant cutting (x⅔).

variety. It is well known, for instance, that the Delaware grape can be propagated more easily in some regions than in others. A common style of single-eye cutting is made with the eye close to the top end, and a naked base of an inch or two. This is inserted into the soil perpendicularly, with the eye just above the surface. It is much used for a variety of plants.

66. Single-eye grape cutting (x½).

Many coniferous plants are increased by cuttings on a large scale, especially retinosporas, arbor-vitæs, and the like. Cuttings are made of the mature wood, which is planted at once (in autumn) in sand under cover, usually in a cool greenhouse (Fig. 67). Most of the species root slowly, and they often remain in the original flats or benches a year, but their treatment is usually simple. In some cases junipers, yews and *Cryptomeria Japonica* will not make roots for nearly twelve months, keeping in good foliage, however, and ultimately giving good plants. They are always grown in shaded houses or frames, and sometimes in inside propagating-frames (Fig. 47). (For more explicit directions, see Thuya and Retinospora, in Chapter VI.)

Most remarkable instances of propagation by means of portions of stems are on record. Chips from a tree trunk have been known to produce plants, and the olive is readily increased by knots or excrescences formed upon the trunks of old trees. These excrescences occur in many plants, and are known as *knaurs*. They are often abundant about the base of large plane-trees, but they are not often used for purposes of propagation. Whole trunks will sometimes grow after having been cut for many months, especially of such plants as

67. Spruce cutting (x½).

cactuses, many euphorbias and yuccas. Sections of these spongy trunks will grow, also. Truncheons of cycad trunks may also give rise to plants (see Chapter VI.). Even saw-logs of our common trees, as elm and ash, will sprout while in the "boom," or water.

Green-wood cuttings are more commonly employed than those from the mature wood, as they "strike" more quickly, they can be handled under glass in the winter, and more species can be propagated by them than by hard-wood cuttings. "Slips" are green-wooded cuttings, but the term is often restricted to designate those which are made by pulling or "slipping" off a small side-shoot, and it is commonly applied to the multiplication of plants in window-gardens. All soft-wooded plants and many ornamental shrubs are increased by green cuttings. There are two general classes of green-wood cuttings: those made from the soft and still growing wood; and those made from the nearly ripened green wood, as in *Azalea Indica*, oleander ficus, etc. House plants, as geraniums, coleuses, carnations, fuchsias, and the like, are grown from the soft young wood, and many harder-wooded plants are grown in the same way. Sometimes true hard wood is used, as in camellia.

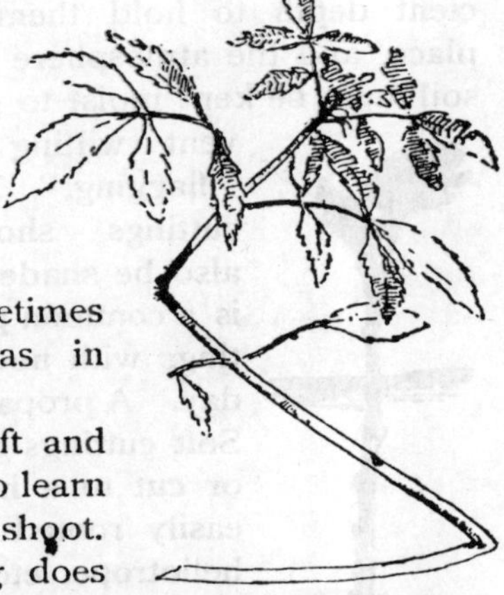

In making cuttings from soft and growing shoots, the first thing to learn is the proper texture or age of shoot. A very soft and flabby cutting does not grow readily, or if it does it is particularly liable to damp-off, and it usually makes a weak plant. Too old wood is slow to root, makes a poor, stunted plant, and is handled with difficulty in many species. The ordinary test for beginners

68. Tough and brittle wood (x½).

is the manner in which the shoot breaks. If, upon being bent, the shoot snaps off squarely so as to hang together with only a bit of bark, as in the upper break in Fig. 68, it is in the proper condition for cuttings; but if it bends or simply crushes, as in the lower portion of the figure, it is either too old or too young for good results.

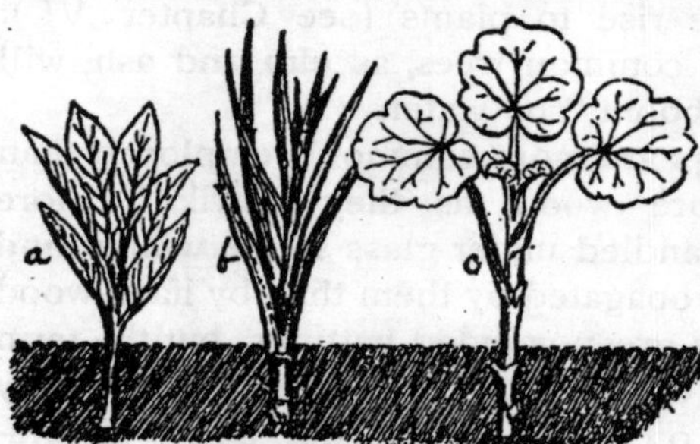

69. Soft cuttings (x⅓).

The tips of the shoots of soft-wooded plants are usually employed, and all or a portion of the leaves are allowed to remain. The cuttings are inserted in sharp sand to a sufficient depth to hold them in place, and the atmosphere and soil must be kept moist to prevent wilting or "flagging." The cuttings should also be shaded for the first week or two. It is a common practice to cover newly set cuttings with newspapers during the heat of the day. A propagating-frame is often employed. Soft cuttings are commonly cut below a bud or cut to a heel, but this is unnecessary in easily rooted plants like geranium, coleus, heliotrope, etc. Fig. 69 shows an oleander cutting at *a*, a carnation at *b*, and a geranium at *c*. A coleus cutting is illustrated in Fig. 70. Many growers prefer to make a larger cutting of some firm-wooded plants, like chrysanthemums, as shown in Fig. 71.

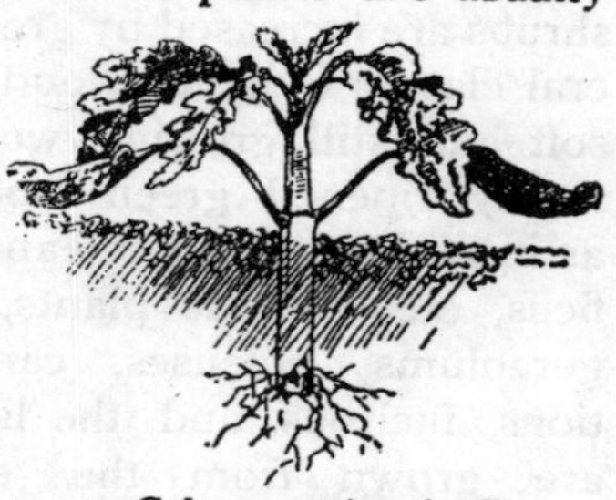

70. Coleus cutting (x⅓).

71. One style of chrysanthemum cutting (x⅓).

Sometimes the growth is so short or the stock so scarce that the cutting cannot be made long enough to hold itself in the soil. In such case a toothpick or splinter is tied to the cutting to hold it erect, as in the cactus cutting, Fig. 72, or the geranium cutting, Fig. 73. In the window garden, soft cuttings may be started in a deep plate which is filled half or two-thirds full of sand and is then filled to the brim with water, and not shaded; this method, practiced on a larger scale, is sometimes useful during the hot summer months. If bottom heat is desired, the plate may be set upon the back part of the kitchen stove. Oleanders usually root best when mature shoots are placed in bottles of water. Refractory subjects may be inserted through the hole in the bottom of an inverted flower-pot, as explained on pages 51 and 52.

72. Cactus cutting held by splinter (x⅓).

Cuttings from the nearly mature green wood are employed for hard-wooded trees and shrubs, as diervillas (weigela), roses, hydrangeas, lilacs, etc. They are cut in essentially the same manner as the hard-wood cuttings described on page 55. They are often taken in summer, when the buds have developed and the wood has about attained its growth. They are cut to two to four or five buds, and are planted an inch or two deep in shaded frames. They are kept close for some days after setting, and the tops are sprinkled frequently. Care must be taken not to set them too deep; they are rarely put in over an inch, if the cutting is six or seven inches long.

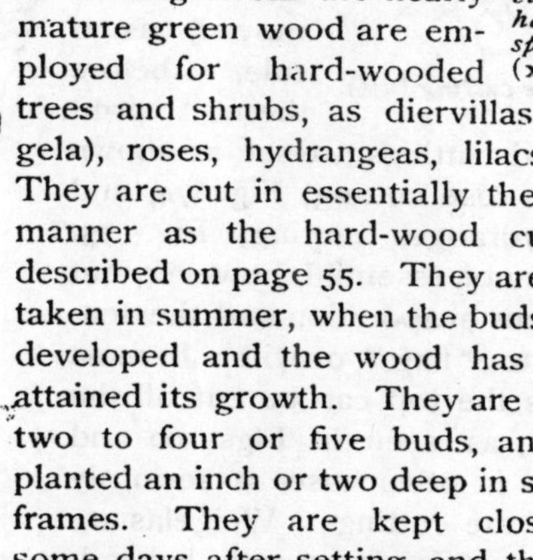

73. Cutting held by toothpick (x⅓).

"June-struck cuttings" are sometimes advantageously

made; here the young shoots of hardy shrubs are taken, when about 2 to 3 inches long, the leaves partly removed, and they are planted under glass. Several weeks are required for rooting, but good plants are obtained, which, when wintered in a coldframe, can be planted out in beds the next spring. Great care must be given to shading and watering. *Hydrangea paniculata* var. *grandiflora* and *Akebia quinata* are examples; or any deutzia or more easily handled plant of which stock is scarce may be cited.

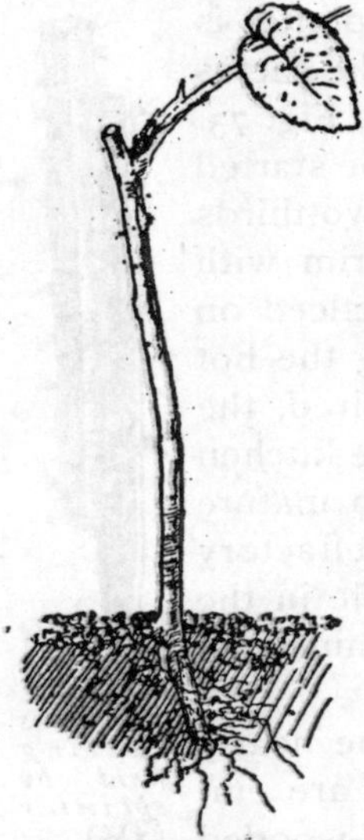

74. Rose cutting (x½).

Part of the leaves are removed, as a rule, before these firm-wooded cuttings are set, as shown in the rose cutting, Fig. 74, and the hydrangea cutting, Fig. 75. This is not essential, however, but it lessens evaporation and the tendency to "flag" or wilt. In most species the top can be cut off the cutting, as seen in Figs. 60 and 75, but in other cases it seriously injures the cutting. Weigelas are likely to suffer from such beheading; an unusually large callus forms at the bottom, but the leaves shrivel and die. This frequently occurs in what some nurserymen call "end growers," among which may be mentioned weigelas (properly diervillas), the shrubby altheas, *Cercis Japonica*, and

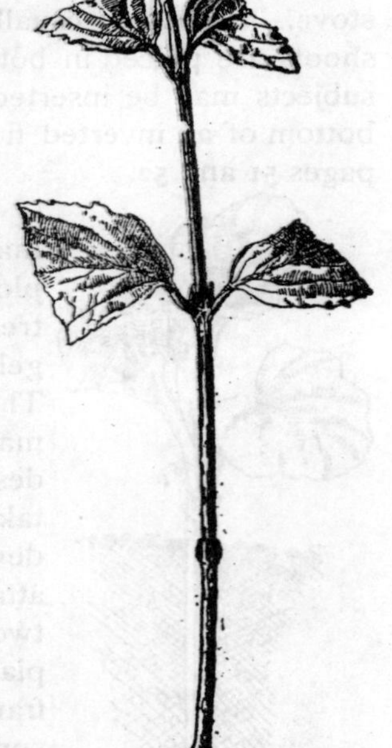

75. Hydrangea cutting (x½).

such spireas as *S. cratægifolia*, *S. rotundifolia* var. *alba*, and *S. Cantonensis* (*S. Reevesii* of the trade), var. *robusta*. The reader must not suppose, however, that all rose cuttings are made after the fashion of Fig. 74, although that is a popular style. Tea roses, and other forced kinds, are very largely propagated from softer wood cut to a single eye, with most or all of the leaf left on (Fig. 76).

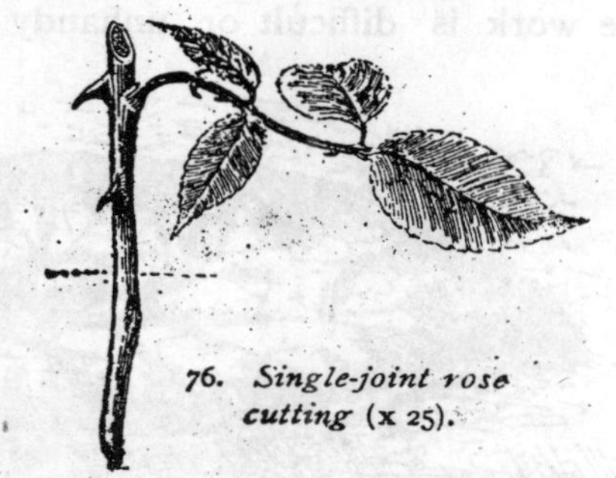

76. *Single-joint rose cutting* (x 25).

These firm-wood cuttings, about two inches long, are often made in the winter from forced plants. Cuttings taken in February, in the north, will be ready to transfer to borders or nursery beds when spring opens. Stout, well-rooted stock-plants are used from which to obtain the cuttings, and they are cut back when taken to the house in the fall, in order to induce a good growth. Many

77. *Young plants from a leaf of Bryophyllum calycinum* (x½).

hardy shrubs can be easily propagated in this way when the work is difficult or unhandy in the open air: *e. g.*,

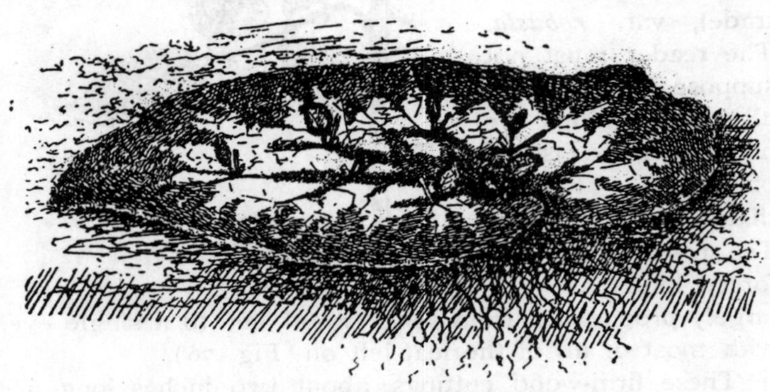

78. Begonia leaf cutting (x½).

Spiræa Cantonensis (*S. Reevesii* of the catalogues), and *S. Van Houttei*, the roses and the like. Stock plants of the soft species, like coleus, lantanas and geraniums, are obtained in like manner.

Leaf Cuttings.—Many thick and heavy leaves may be used as cuttings. Leaf cuttings are most commonly employed in the showy-leaved begonias, in succulents, and in gloxinias, but many plants can be propagated by them. Even the cabbage can be made to grow from leaf cuttings. The bryophyllum is one of the best plants for showing the possibilities of propagation by leaves. If one of the thick leaves is laid upon moss or sand in a moist atmosphere, a young plant will start from nearly every pronounced angle in the margin (Fig. 77). In Rex begonias, also, the

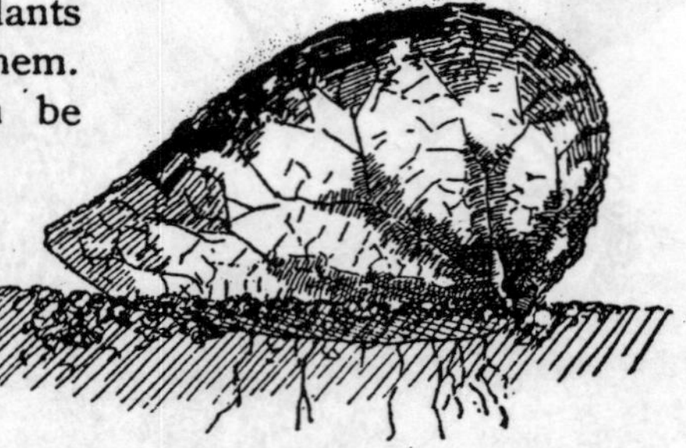

79. An upright begonia leaf cutting (x⅓).

whole leaf may be used, as shown in Fig. 78. It is simply laid upon moist sand in a frame and held down by splinters thrust through the ribs. The wound made by the peg induces the formation of roots, and a young plant arises. A half dozen or more plants can be obtained from one leaf. Some operators cut off the ribs, instead of wounding them with a prick. Many gardeners prefer to divide the leaf into two nearly equal parts, and then set each part, or the better one, upright in the soil, the severed edge being covered. This is shown in Fig. 79. Fewer plants—often only one—are obtained in this manner, but they are strong.

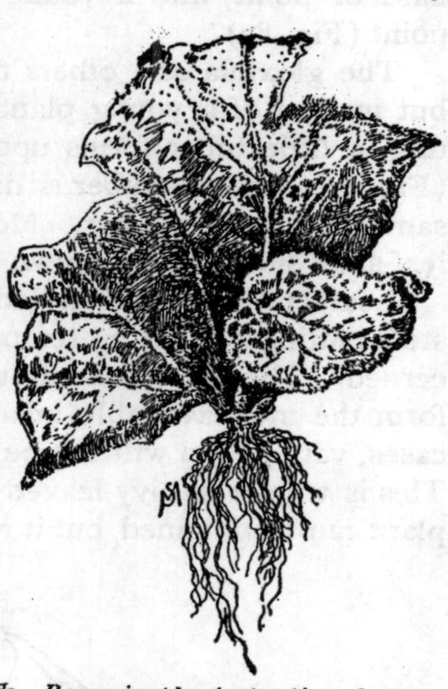

80. Begonia plant starting from a triangular leaf cutting (x¾).

When stock is scarce the begonia leaf may be cut into several fan-shaped pieces. The whole leaf may be divided into as many triangular portions as can be secured with a portion of the petiole, a strong rib, or a vein attached at the base; these pieces, inserted and treated like coleus cuttings, will root and make good plants within a reasonable time, say six months. This form of cutting should be two to three inches long by an inch or inch and one-half wide. Ordinarily, in this style of leaf cutting, the petiole or stalk is cut off close to the leaf and the lower third or fourth of the leaf is then cut off by a nearly straight cut across the leaf. This somewhat triangular base is then cut into as many wedge-shaped pieces as there are ribs in the leaf, each rib forming the center of a cutting. The point of each cutting should contain a portion of the petiole. The points of these triangular portions are inserted in the soil a half inch or so, the cutting standing erect or nearly so. Roots form at the base or point, and a young plant springs from the same point (Fig. 80).

The gloxinia and others of its kin propagate by leaves, but instead of a young plant arising directly from the cutting, a little tuber forms upon the free end of the petiole (Fig. 81), and this tuber is dried off and finally planted the same as a mature tuber. Most gardeners prefer to cut the leaf-stalk shorter than shown in the cut.

Leaf cuttings are handled in the same manner as soft stem-cuttings, so far as temperature and moisture are concerned. There are comparatively few species in which they form the most available means of multiplication. In some cases, variegation will not be reproduced by the rooted leaf. This is true in the ivy-leaved geranium L'Elegante ; a good plant can be obtained, but it reverts to the plain-leaved type.

81. Leaf cutting of gloxinia (x¼).

CHAPTER V.

GRAFTAGE.

I. GENERAL CONSIDERATIONS.

GRAFTAGE is rarely employed for the propagation of the species itself, as seedage and cuttage are more expeditious and cheaper. Its chief use is to perpetuate a variety which does not reproduce itself from seeds, and which cannot be economically grown from cuttings. Graftage is always a secondary operation. That is, the root or stock must first be grown from seeds, layers or cuttings, and this stock is then grafted or budded to the desired variety. Graftage is employed in the propagation of every kind of tree fruits in America, and of very many ornamental trees and shrubs, and it is indispensable to the nursery business. It therefore needs to be discussed in considerable detail.

All the named varieties of tree fruits and many of those of ornamental trees and shrubs are perpetuated by means of graftage. In some species, which present no marked or named varieties, however, propagation by seeds or cuttings is for various reasons so difficult or uncertain that recourse must be had to graftage. This is particularly true in many of the firs and spruces, which do not produce seeds to any extent in cultivation. In other cases, graftage is performed for the purpose of producing some radical change in the character or habit of the plant, as in the dwarfing of pears by grafting them upon the quince, and of apples by grafting on the Paradise stock, the elevation of weeping tops by working them upon upright trunks, and the acceleration of fruit-bearing by setting cions in old plants. It is sometimes employed to aid the healing of wounds or to repair and fill out broken tops. And it has been used to make infertile plants fertile, by grafting in the missing sex in diœcious trees, or a variety with more potent pollen, as practiced in some of the native plums. All these uses of graftage fall under three heads: 1. To perpetuate a variety. 2. To increase the ease and speed of multiplication. 3. To produce some radical change in nature or habit of cion or stock.

Mutual Influence of Stock and Cion.—The first two divisions in the above paragraph need no elaboration here, but the third is moot ground, and demands subdivision. These secondary results of grafting, as they may be called, or reciprocal influences of stock and cion, fall readily under the following heads (which were outlined by the writer in Garden and Forest for February 26, 1890):

1. Graftage may modify the stature of the plant. It is the commonest means of dwarfing plants. We graft the pear upon the quince and the apple upon the Paradise apple. This dwarfing usually augments proportionate fruitfulness. (For further discussion of dwarfing, see page 147.)

2. Graftage may be made the means of adapting plants to adverse soils. Illustrations are numerous. Many varieties of plums, when worked on the peach, thrive in light soils, where plums on their own roots are uncertain. Conversely, some peaches can be adapted to heavy soils by working on the plum. If dwarf pears are desired on light soils, where the quince does not thrive, recourse is had to grafting on the mountain ash, or some of its allies. In some chalky districts of England the peach is worked on the almond. Some plums can be grown on uncongenial loose soils by working them on the Beach plum. Professor Budd states, in Garden and Forest for February 12, 1890, that the Gros Pomier apple is particularly adapted to sandy land and the Tetofsky to low prairie land, and that these stocks are often selected to overcome adversities of soil. Such instances are frequent, and should demand greater attention from cultivators.

3. Graftage may be made the means of adapting plants to adverse climate. This may be brought about by either or both of two causes: (*a*) The early maturation of the stock, causing the cion to ripen better. The Oldenburgh apple is a favorite stock in severe climates for this reason. The Siberian crab often has the same influence, although its use may be open to objection. (*b*) The mechanical effect of the union, impeding the passage of sap and causing the cion to mature or ripen early. This fact has been observed in many cases, notably in some instances of apples upon improved crabs, and yet the union is perfect enough, nevertheless, to maintain the plant in a profitable condition for years. There are some apparent adaptations to climate, however, which are not explained by either of the above hypotheses.

4. Graftage may correct a poor habit. All propagators are aware of this fact. The Canada Red apple is usually top-worked to overcome its weak and straggling habit. The Winter Nelis pear is a familiar example.

5. Graftage is often the means of accelerating fruitfulness: *i. e.*, plants are made to bear at an earlier age. Those who test new orchard fruits are familiar with this fact. Cions from young trees bear sooner if set in old trees than when set in young ones. This result may sometimes be due to the same causes which abbreviate the vigor of plants, as already outlined (see § 3, above). Checking growth induces fruitfulness.

6. Graftage often modifies the season of ripening of fruit. This is brought about by different habits of maturity of growth in the stock and cion. An experiment with Winter Nelis pear showed that fruit kept longer when grown upon Bloodgood stocks than when grown upon Flemish Beauty stocks. The latter stocks in this case evidently completed their growth sooner than the others. Twenty Ounce apple has been known to ripen in advance of its season by being worked upon Early Harvest. Mr. Augur cites an instance in which the Roxbury Russet,

grafted upon the Golden Sweet, which is early in ripening, was modified both in flavor and keeping qualities. "Keeping qualities" is but another expression for "season of ripening." These influences are frequent; in fact, they are probably much commoner than we are aware.

7. Graftage often augments fruitfulness, largely for the same reasons as discussed in § 3. There are some anomalous instances of increase of fruitfulness which are difficult of explanation; *e. g.*, some citrus fruits are more productive when grafted upon *Citrus* (or *Ægle*) *trifoliata* than upon their own roots.

8. Graftage often delays the degeneration of varieties. In various ornamental plants this influence is marked, as compared with plants from cuttings. It is recorded particularly in certain roses and camellias.

9. Graftage sometimes increases the size of fruit. The best illustrations of this fact are found in certain pears when grown upon the quince; the fruit is often larger than from standard trees.

10. Graftage may result in a modification of color of foliage, flowers or fruit. Assumed influences of this character are frequently recorded, but it is not always possible to determine how much of the modification may be due to soil, climate and treatment. *Prunus Pissardii* has been seen to give much more highly colored foliage when grafted upon *Prunus Americana* than upon *P. domestica*. The cions came from the same tree, and the grafted trees stood in the same row. Any acceleration in ripening of fruit (as indicated in § 5) is apt to cause high color, but the intensification of color in *Prunus Pissardii* was not due to such cause, as the grafts were more vigorous upon *P. Americana*.

11. Graftage may influence the flavor of fruit. There can be no question but that apples often derive acridity from the stock when worked upon the wild crab or upon the Siberian crab. It is said that the Angoulême and some other pears are improved in flavor when grown upon

the quince. Downing asserts that some varieties "are considerably improved in flavor" by working upon quince. Similar results may occur in the dwarfing of apples.

A favorite illustration in support of the reciprocal influences of stock and cion is the fact of transfer of color or variegation by grafting. Darwin called attention to this phenomenon, and used the term "graft-hybrid" to designate similar mongrel offspring of certain unions. But this class of phenomena seems to follow inoculation rather than grafting *per se*. The transferable nature of variegation is well known in certain species, but it is entirely inexplicable in the present state of our knowledge; it seems certain, however, that it does not merit attention under a discussion of grafting. So long ago as 1727 variegation was designated by Bradley a "distemper," which "may be communicated to every plant of the same tribe by inoculating only a single bud." In our own day, Morren has called it the "contagion of variegation."

The above outline illustrates the fact that the results of graftage are profoundly modified by conditions. Adverse conditions must give unsatisfactory results, and may lead to a premature denunciation of the whole system of propagation upon the roots of other plants. But, on the other hand, proper conditions and good execution afford abundant and positive proof that graftage is essential to best success in many departments of horticulture.

Limits of Graftage.—Probably all exogenous plants—those which possess a distinct bark and pith—can be grafted. Plants must be more or less closely related to each other to allow of successful graftage of the one upon the other. What the affinities are in any case can be known only by experiment. As a rule, plants of close botanical relationship especially those of the same genus, graft upon each other with more or less ease; yet this relationship is by no means a safe guide. A plant will often thrive better upon a species of another genus than upon a congener. The pear, for instance, does better upon many thorns than upon the apple. Sometimes plants of very distinct genera unite readily. Thus among cacti, the leafless epiphyllum grows remarkably well upon the leaf-bearing pereskia. It should be borne in mind that union of tissues is not a proof of affinity. Affinity can be measured only by the thrift, healthfulness and longevity of the cion. The bean has been known to make a union with the chrysanthemum, but it almost immediately died. Soft tissues, in particular, often combine in plants which possess no affinity whatever, as we commonly understand the term. Neither does affinity refer to relative sizes or rates of growth of stock and cion, although the term is sometimes used in this sense. It cannot be said that some varieties of pear lack affinity for the quince, and yet the pear cion grows much larger than the stock. In fact, it is just this difference in size and rate of growth which constitutes the value of the quince root for dwarfing the pear. When there is a marked difference in rate of growth between the stock and cion, an enlargement will occur in the course of time, either above or below the union. If this occurs upon the stem, it makes an unsightly tree. If the cion greatly outgrows the stock a weak tree is the result.

General Methods.—Graftage can be performed at almost any time of the year, but the practice must be greatly varied to suit the season and other conditions. The one essential point is to make sure that the cambium layers, lying between the bark and wood, meet as nearly as possible in the cion and stock. This cambium is always present in live parts, forming woody substance from its inner surface and bark from its outer surface. During the season of greatest growth it usually occurs as a soft, mucilaginous and more or less unorganized substance, and in this stage it most readily repairs and unites wounded surfaces. And for this reason the grafting and budding of old trees are usually performed in the spring. Later in the season, the cambium becomes firmer and more differentiated, and union of woody parts is more uncertain. It is also necessary to cover the wounds in order to check evaporation from the tissues. In outdoor work wax is commonly used for all species of graftage which wound the wood itself, but in budding, the loosened bark, bound down securely by a bandage, affords sufficient protection. It is commonly supposed that an ordinary cleft-graft cannot live if the bark of the stock immediately adjoining it is seriously wounded, but the bark really serves little purpose beyond protection of the tissues beneath. A cion will grow when the bark is almost entirely removed from the stub, if some adequate protection can be given which will not interfere with the formation of new bark. The cion must always possess at least one good bud. In most cases, only buds which are mature or nearly so are used, but in the grafting of herbs very young buds may be employed. These simple requirements can be satisfied in an almost innumerable variety of ways. The cion or bud may be inserted in the root, crown, trunk, or any of the branches; it may be simply set under the bark, or inserted into the wood itself in almost any fashion; and the operation may be performed either upon growing or dormant plants at any season. But in practice there are comparatively few methods which are sufficiently simple and expeditious to admit of indiscriminate use; the operator must be able to choose the particular method which is best adapted to the case in hand.

Classification of Graftage. — There are three general divisions or kinds of graftage, between which, however, there are no decisive lines of separation: 1. Bud-grafting, or budding, in which a single bud is inserted upon the surface of the wood of the stock. 2. Cion-grafting, or grafting proper, in which a detached twig, bearing one or more buds, is inserted into or upon the stock. 3. Inarching, or grafting by approach, in which the cion remains attached to the parent plant until union takes place. This last is so much like grafting proper, and is so little used, that it is discussed under the head of grafting in the succeeding parts of this chapter. Each of these divisions can be almost endlessly varied and sub-divided, but in this discussion only the leading practices can be detailed. The following enumeration, after Baltet, will give a fair idea of the kinds of grafting which have been employed under distinct names:

1. Bud-Grafting, or Budding.

1.—Grafting with shield-buds.
Bud-grafting under the bark, or by inoculation.
" " ordinary method.
" " with a cross-shaped incision.
" " " the incision reversed.
" " by veneering.
Bud-grafting, the combined or double method.
2.—Flute-grafting.
" " common method.
" " with strips of bark.

2. Cion-Grafting, or Grafting proper.

1.—Side-grafting under the bark.
" " with a simple branch.
" " with a heeled branch.
" " in the alburnum.
" " with a straight cleft.
" " with an oblique cleft.
2.—Crown-grafting.
Ordinary method.
Improved method.
3.—Grafting *de precision.*
Veneering, common method.
" in crown-grafting.
" with strips of bark.
Crown-grafting by inlaying.
Side-grafting by inlaying.

4.—Cleft-grafting, common single.
" " common double.
" " oblique.
" " terminal.
" " " woody.
" " " herbaceous.

5.—Whip-grafting, simple.
" " complex.
Saddle-grafting.

6.—Mixed grafting.
Grafting with cuttings.
When the cion is a cutting.
When the stock is a cutting.
When both are cuttings.
Root-grafting of a plant on its own root.
" " " " the roots of another plant.
Grafting with fruit buds.

3. Inarching, or Grafting by Approach.

1.—Method by veneering.
" " inlaying.
English method.

2.—Inarching with an eye.
" " a branch.

Is Graftage a Devitalizing Process?—The opinion is commonly expressed by horticultural writers that graftage is somehow vitally pernicious, and that its effects upon the plant must be injurious. Graftage is often cited as the cause of the running out of varieties. The process has also been strongly indicted during the past few years by writers in England Inasmuch as the question is vital to the practice of fruit-growing in America, it will be worth while to make a somewhat careful study of the questions respecting the relationship between graftage and the vitality of the grafted plant. For this purpose, a paper read by the author before the Peninsula Horticultural Society at Dover, Delaware, in 1892, and printed in the transactions of the society, is here reproduced:

To the popular mind there seems to be something mysterious in the process of graftage. People look upon it as something akin to magic, and entirely opposed to the laws of nature. It is popularly thought to represent the extreme power which man exercises over natural forces. It is strange that this opinion should prevail in these times, for the operation itself is very simple, and the process of union is nothing more than the healing of a wound. It is in no way more mysterious than the rooting of cuttings, and it is not so unnatural, if by this expression we refer to the relative frequency of the occurrences of the phenomena in nature. Natural grafts are by no means rare among forest trees, and occasionally the union is so complete that the foster stock entirely supports and nourishes the other. A perfect inarch-graft, by means of which two oak trees have united into one, is shown in Fig. 82. Cuttings of stems, however, are very rare among wild plants; in fact, there is but one common instance, in the north, in which stem cuttings are made entirely without the aid of man, and that is the case of certain brittle willows whose branchlets are easily cast by wind and snow into streams and moist places, where they sometimes take root. But mere unnaturalness of any operation has no importance in discussions of phenomena attaching

82. A natural graft of forest trees.

to cultivated plants, for all cultivation is itself unnatural in this ordinary sense.

But it is difficult to see why the union of cion and stock is any more mysterious or unusual than the rooting of cuttings; in fact, it has always seemed to me to be the simpler and more normal process of the two. A wounded surface heals over as a matter of protection to the plant, and when two wounded surfaces of consanguineous plants are closely applied, nothing is more natural than that the nascent cells should interlock and unite. In other words, there is no apparent reason why two cells from different allied stems should refuse to unite any more than two cells from the same stem. But why bits of stem should throw out roots from their lower portion and leaves from their upper portion, when both ends may be to every human sense exactly alike, is indeed a mystery. Healing is regarded as one of the necessary functions of stems, but rooting cannot be so considered.

This much is said by way of preface in order to eliminate any preconception that graftage is in principle and essence opposed to nature, and is therefore fundamentally wrong. A large part of the discussion of the philosophy of grafting appears to have been random, because of a conviction or assumption that it is necessarily opposed to natural processes.

It does not follow from these propositions, however, that graftage is a desirable method of multiplying plants, but simply that the subject must be approached by means of direct and positive evidence. Much has been said during the last few years concerning the merits of graftage, and the opponents of the system have made the most sweeping statements of its perniciousness. This recent discussion started from an editorial which appeared in The Field, an English journal, and which was copied in The Garden of January 26, 1889, with an invitation for discussion of the subject. The article opens as follows: "We doubt if there is a greater nuisance in the whole practice of gardening than the art of grafting. It is very clever, it is very interesting, but it will be no great loss if it is abolished altogether. It is for the convenience of the nurseryman that it is done in nine cases out of ten, and in nearly all instances it is not only needless, but harmful. * * * If we made the nurserymen give us things on their own roots, they would find some quick means of doing so." A most profuse discussion followed for a period of two years, in which many excellent observers took part. Some of the denunciations of graftage are as follows: "Grafting is always a makeshift, and very often a fraud." "Grafting is in effect a kind of adulteration. * * * It is an analogue of the coffee and chicory business. Grafted plants of all kinds are open to all sorts of accidents and disaster, and very often the soil, or the climate, or the cultivator, is blamed by employers for evils which thus originated in the nursery. * * * If, in certain cases, grafting as a convenience has to be resorted to, then let it be root-grafting, a system that eventually affords the cion a chance of rooting on its own account in a natural way." "Toy games, such as grafting and budding, will have to be abandoned, and real work must be begun on some sound and sensible plan." "Any fruit-bearing or ornamental tree that will not succeed on its own roots had better go to the rubbish fire at once. We want no coddled or grafted stuff, when own-rooted things are in all ways infinitely better, healthier, and longer-lived." These sweeping statements are made by F. W. Burbidge, of Dublin, a well-known author, whose opinions command attention. The editor of The Garden writes: "We should not plant any grafted tree or shrub, so far as what are called ornamental trees and shrubs are concerned. There may be reason for the universal grafting of fruit trees, though we doubt it." These quotations are not cited in any controversial spirit, but simply to show the positiveness with which the practice of graftage is assailed. As the presumption is in favor of any practice which has become universal, these statements possess extraordinary interest.

The assumptions underlying these denunciations of graftage are three, and as these are essentially the reasons which are usually cited by the opponents of the system, they may be considered here. These are: 1. The citation of numerous instances in which graftage (by which is meant both grafting and budding) has given pernicious results. 2. The affirmation that the process is unnatural. 3. The statement that own-rooted plants are better — that is, longer lived, hardier, more virile — than graft-rooted plants.

1. The citations of the injurious effects of graftage are usually confined to ornamental plants, and the commonly cited fault of the operation is the tendency of the stocks to sucker and choke the graft. This fault is certainly very common, but on the other hand there are numerous instances in which it does not occur, as, for instance, in peach, apple, pear and many other fruit-trees, and in very many ornamentals. In fact, it is probably of no more common occurrence than is the pernicious suckering of plants grown from cuttings, as in the lilacs, cutting-grown or sucker-grown plums, and many other plants, in which suckers must be assiduously kept down or they will choke the main stem which we are endeavoring to rear. And these remarks will apply with equal force to all the citations of the ill-effects of graftage; the cases simply show that the operation has been a failure or is open to objections in the particular instances cited, and they afford no proof that there may not be other plants upon which graftage is an entire success. Graftage has been indiscriminately employed, and it is apparent to everyone that there have been many failures. But this does not prove graftage wrong, any more than the wrong practice of physicians proves that the science of medicine is pernicious. If there are plants upon which graftage is entirely successful, then all must agree that the operation itself,

per se, is not wrong, however many cases there may be to which it is not adapted.

2. The proposition that graftage is unnatural, and therefore pernicious, is no more nor less than a fallacy. In the first place, there is nothing to show that it is any more unnatural than the making of cuttings, and if naturalness is proved by frequency of occurrence in nature, then graftage must be considered the more natural process of the two, as already shown. One of the most determined writers upon this subject has said that "it is quite fair to say that raising a tree from seed, or a shrub by pulling it in pieces [cuttings] is a more natural mode of increase than by grafting." It is difficult to understand by what token the author is to prove that pulling a plant in pieces is more natural than graftage; and there appears to have been no attempt to show that it is so.

But the whole discussion of the mere naturalness of any operation is really aside from the question, for every operation in the garden is in some sense unnatural, whether it be transplantation, pruning, or tillage; and it is well known that these unnatural processes may sometimes increase the longevity and virility of the plant. Plants which are given an abundance of food and are protected from insects and fungi and the struggle with other plants, are better equipped than those left entirely to nature. It is the commonest notion that cultivation is essentially an artificial stimulus, that it excites the plant to performances really beyond its own power, and therefore devitalizes it. But this is a fallacy. All plants and animals in a state of nature possess more power than they are able to express, and they are held in a state of equilibrium, as Herbert Spencer puts it, by the adaptation to environment. Once the pressure of existing environments is removed, the plant springs into the breach and takes on some new features of size, robustness, or prolificacy, or distributes itself in new directions. The whole series of benefits which arise from a change of seed is a familiar proof of this fact. So that, if cultivation, domestication, or, in other words, unnaturalness, may be sometimes a stimulus, it is not necessarily so. Cultivation differs from natural conditions more in degree than in kind. Or, as Darwin writes, "Man may be said to have been trying an experiment on a gigantic scale," and "it is an experiment which nature during the long lapse of time has incessantly tried."

3. It is said that own-rooted plants are better than foster-rooted ones. This is merely an assumption, and yet it has been held with dogmatic positiveness by many writers. If mere unnaturalness, that is, rarity or lack of occurrence in nature, is no proof of perniciousness, as has been shown, then this statement admits of argument just as much as any other proposition. And surely at this day we should test such statements by direct evidence rather than by *a priori* convictions. The citation of any number of instances of the ill-effects of graftage is no proof that own-rooted plants are necessarily better, if there should still remain cases in which no injurious effects follow. Now, if it is true that "own-rooted things are in all ways infinitely better, healthier and longer-lived" than foster-rooted plants, and if "grafted plants of all kinds are open to all sorts of accidents and disaster," then the proposition should admit of most abundant proof. The subject may be analyzed by discussing the following questions: *a.* Is the union always imperfect? *b.* Are grafted plants less virile than own-rooted ones? Are they shorter lived?

a. It is well known that the physical union between cion and stock is often imperfect, and remains a point of weakness throughout the life of a plant. But this is not always true. There are scores of plants which make perfect physical unions with other plants of their own species, or even with other species, and it follows that these, alone, are the plants that should be grafted. The very best proof which can be adduced that the union may be physically perfect, is to be found in the micro-photograph of an apple graft published six years ago in The American Garden by Professor C. S. Crandall. The cells are knit together so completely that it is impossible to determine the exact line of union (Fig. 83). Mr. Crandall also fig-

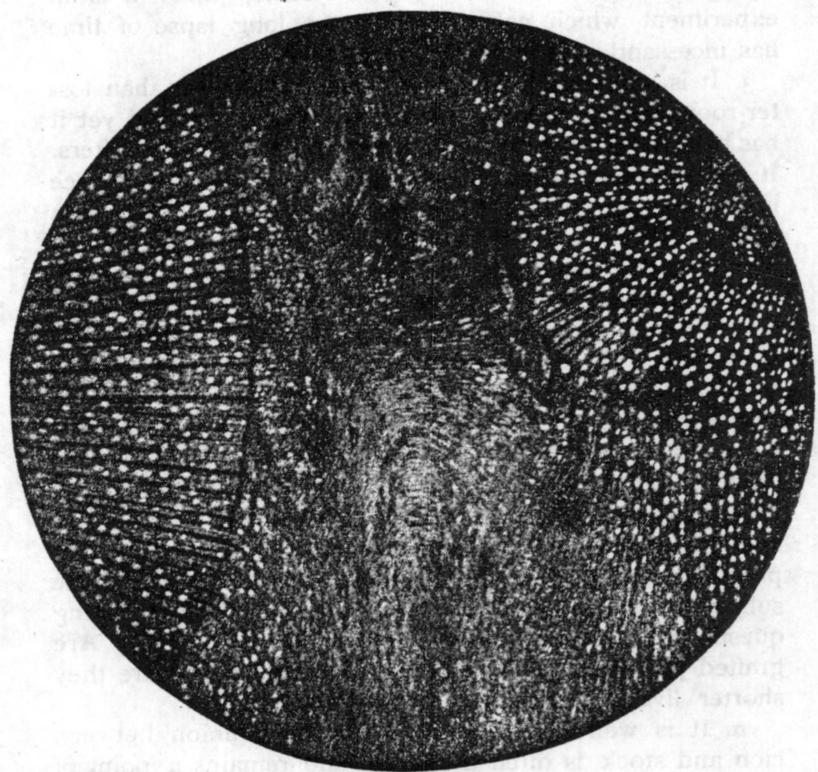

83. A perfect union of stock and cion, following veneer-grafting. The stock is upon the left, and the cion upon the right. The united tissue is seen running through the center. (x 25.)

ures a microscopic section of an apple graft in which the union is very poor, but this graft was made in a different manner from the other (Fig. 84); and that is another proof that the operation should be suited to the subject.

These were grafts made upon nursery stock, and it

would appear that if the union were good at the expiration of the first year, it would remain good throughout the life of the plant. In order to test this point, two apple trees were procured, which were fifteen years old and over

84. Imperfect union of a whip-graft. The body of the stock is on the left; then follows the tongue of the cion; then the tongue of the stock; and finally, on the right, the body of the cion. The spaces show the lack of union which sometimes occurs. (x 25.)

six inches in diameter, which had been grafted at the surface of the ground in the nursery. In the presence of two critical observers, the trunks were split into many pieces, but no mark whatever could be found of the old union. The grain was perfectly straight and bright through the crown. Every internal evidence of a graft had disappeared.

So far as the strength of a good union is concerned, all fruit growers know that trees rarely break where they are grafted. In a certain seedling orchard, many hundreds of grafts were set in the tops of the trees, often far out on large limbs; and yet, with all the breaking of the trees by ice, storms and loads of fruit, a well-established union has not been known to break away. The strength of the union was tested in a different way. Two "stubs" were cut from an old and rather weak apple tree which had been cleft-grafted in the spring of 1889. These stubs were sawed up into cross-sections less than an inch thick, and each section, therefore, had a portion of foreign wood grown into either side of it. These sections were now placed on a furnace and kept very hot for two days, in order to determine how they would check in seasoning, for it is evident that the checks occur in the weakest points. But in no case was there a check in the amalgamated tissue, showing that it was really an element of physical strength to the plant. A similar test was made with yearling mulberry grafts, and with similar results; and this case is particularly interesting because there were three species engrafted—the common Russian mulberry, *Morus rubra*, and *M. Japonica*.

From all these considerations, it is evident that, admitting that hundreds of poor unions occur, there is no necessary reason why a graft should be a point of physical weakness, and that the statement that "grafted plants of all kinds are open to all sorts of accidents and disaster," is not true.

b. Are grafted plants less virile—that is, less strong, vigorous, hardy, shorter-lived—than others? It is evident that a poor union or an uncongenial stock will make the resulting plant weak, and this is a further proof that indiscriminate graftage is to be discouraged. But these facts do not affirm the question. There are two ways of approaching the general question, by philosophical considerations and by direct evidence.

It is held by many persons that any asexual propagation is in the end devitalizing, since the legitimate method of propagation is by means of seeds. This notion appears to have found confirmation in the conclusions of Darwin and his followers, that the ultimate function of sex is to revitalize and strengthen the offspring following the union of the characters or powers of two parents; for if the expensive sexual propagation invigorates the type, asexual propagation would seem to weaken it. It does not follow, however, that because sexual reproduction is good, asexual increase is bad, but rather that the one is, as a rule, better than the other, without saying that the other is injurious. We are not surprised to find, therefore, that some plants have been asexually propagated for centuries with apparently no decrease of vitality, although this fact does not prove that the plant may not have positively increased in virility if sexual propagation had been employed. The presumption is always in favor of sexual reproduction, a point which will be admitted by every one. And right here is where graftage has an enormous theoretical advantage over cuttage or any other asexual multiplication: the root of the grafted plant springs from sexual reproduction, for it is a seedling, and if the union is physically perfect, as is frequently the case, there is reason to suppose that grafting between consanguineous plants is better than propagating by cuttings or layers. In other words, graftage is really sexual multiplication, and if seeds have any advantage over buds in forming the foundation of a plant, graftage is a more perfect method than any other artificial practice. It is, in fact, the nearest approach to direct sexual reproduction, and when seeds cannot be relied upon wholly, as they cannot, for the reproduction of many garden varieties, it is the ideal practice, always provided, of course, that it is properly done between congenial subjects. It is not to be expected that the practice is adapted to all plants, any more than is the making of cuttings of leaves or of stems, but this fact cannot be held to invalidate the system.

It has been said, in evidence that graftage is a devitalizing or at least disturbing process, that grafted plants lose the power of independent propagation. Mr. Burbidge writes that "any plant once grafted becomes exceedingly difficult of increase, except by grafting." Evidence should be collected to show if this is true. All our fruits grow just as readily from seeds from grafted as from seedling trees, and it is doubtful if there is a well authenticated case of a plant which grows readily from cuttings becoming any more difficult to root from cuttings after having been grafted.

But is there direct evidence to show that "grafting is always a make-shift," that it is a "toy game," that "grafted plants of all kinds are open to all sorts of accidents and disaster," that "own-rooted things are in all ways infinitely better, healthier, and longer-lived?" These statements allow of no exceptions; they are universal and iron-bound. If the questions were to be fully met, we should need to discuss the whole art of graftage in all its detail, but if there is one well authenticated case in which a grafted plant is as strong, as hardy, as vigorous, as productive and as long-lived as seedlings or as cutting-plants, we shall have established the fact that the operation is not necessarily pernicious, and shall have created the presumption that other cases must exist.

Some forty years ago, a traveller took apple seeds from his old home in Vermont and planted them in Michigan. The seeds produced some hundred or more lusty trees, but as most of the fruit was poor or indifferent, it was decided to top-graft the trees. This grafting was done in the most desultory manner, some trees being grafted piece-meal, with some of the original branches allowed to remain permanently, while others were entirely changed over at once;

and a few of them had been grafted on the trunk about three or four feet high, when they were as large as broomsticks, the whole top having been cut off when the operation was performed. A few trees which chanced to bear tolerable fruit, scattered here and there through the orchard, were not grafted. The orchard has been, therefore, an excellent experiment in grafting. Many of the trees in this old orchard have died from undeterminable causes, and it is an interesting fact that fully half, and probably even more, of the deaths have been seedling trees which were for many years just as vigorous in every way as the grafted trees; and of the trees that remain, the grafted specimens are in every way as vigorous, hardy and productive as the others. Some of these trees have two tops, one of which was grafted shoulder high in the early days, and the other grafted into the resulting top many years later. And those trees which contain both original branches and grafted ones in the same top show similar results—the foreign branches are in every way as vigorous, virile and productive as the others, and they are proving to be just as long-lived. Here, then, is a positive experiment compassed by the lifetime of one man, which shows that own-rooted trees are not always "infinitely better, healthier, and longer-lived" than grafted plants. This illustration may be considered as a type of thousands of orchards, containing various fruits, in all parts of the country. The fact may be cited that the old seedling orchards which still remain to us about the country are much more uneven and contain more dead trees or vacant places than the commercial grafted orchards of even the same age. This is due to the struggle for existence in the old orchards, by which the weak trees have disappeared, while the grafted orchards, being made up of selected varieties of known virility and hardiness, have remained more nearly intact, and if the seedling orchards have suffered more than the grafted ones, it must be because they have had more weak spots.

The universal favor in which graftage is held in America is itself a strong presumption in its favor. Growers differ among themselves as to the best methods of performing the operation, but an intelligent American will not condemn the system as necessarily bad or wrong. In 1890 there were growing in the United States nurseries 240,570,666 apple trees, 88,494,367 plum trees, 77,223,402 pear trees, and 49,887,874 peach trees, with enough other species to make the total of fruit trees 518,016,612. All of this vast number will go as grafted or budded trees to the consumer, and he will accept none other. It is true that half of them may die from various causes before they reach bearing age, but graftage itself plays a small part in the failure, as may be seen in the case of grapes and small fruits, which outnumber the tree fruits in nursery stock, and of which less than one-half probably reach maturity, and yet these are cutting-grown plants. It is, in nineteen cases out of twenty, the carelessness of the grower which brings failure.

It is impossible, if one considers the facts broadly and candidly, to arrive at any other conclusion than this: Graftage is not suited to all plants, but in those to which it is adapted—and they are many—it is not a devitalizing process.

2. BUDDING.

Budding is the operation of applying a single bud, bearing little or no wood, to the surface of the growing wood of the stock. The bud is applied directly to the cambium layer of the stock. It is nearly always inserted under the bark of the stock, but in flute-budding a piece of bark is entirely removed, and the bud is used to cover the wound. There is no general rule to determine what species of plants should be budded and which ones cion-grafted. In fact, the same species is often multiplied by both operations. Plants with thin bark and an abundance of sap are likely to do best when grafted; or if they are budded, the buds should be inserted at a season when the sap is least abundant, to prevent the "strangulation" or "throwing out" of the bud. In such species, the bark is not strong enough to hold the bud firmly until it unites; and solid union does not take place until the flow of sap lessens. Budding is largely employed upon nearly all young fruit trees, and almost universally so upon the stone fruits. It is also used in roses and many ornamental trees. Upon nursery trees, it is employed in a greater number of cases than grafting is, but grafting is in commoner use for working-over the tops of large trees. Budding is commonly performed during the growing season, usually in late summer or early fall, because mature buds can be procured at that time, and young stocks are then large enough to be worked readily. But budding can be done in early spring, just as soon as the bark loosens; in this case perfectly dormant buds must have been taken in winter and kept in a cellar, ice-house or other cool place. Budding is always best performed when the bark slips or peels easily. It can be done when the bark is tight, but the operation is then tedious and uncertain. It is also much more successful when performed in dry, clear weather.

85. *Shield-bud* (x 1).

Shield-budding.—There is but one style of budding in general use in this country. This is known as shield-budding, from the shield-like shape of the portion of bark which is removed with the bud. Technically, the entire severed portion, comprising both bark and bud, is called a "bud" A shield-bud is shown natural size in Fig. 85. This is cut from a young twig of the present season's growth. It is inserted underneath the bark of a young stock or branch (Fig. 91), and is then securely tied, as shown in Fig. 92.

The minor details of shield-budding differ with nearly every operator, and with the kind of plant which is to be budded. In commercial practice, it is performed in the north mostly from early July until the middle of September. In the southern states it usually begins in June. As a rule, apples and pears are budded earlier in the season than peaches are. This is due to the fact that peach stocks are nearly always budded the same season the pits are planted, and the operation must be delayed until the stocks are large enough to be worked.

Most fruit-stocks, especially apples and pears, are not budded until two years after the seeds are sown. The plants grow for the first season in a seed-bed. The next spring they are transplanted into nursery rows, and budded when they become large enough, which is usually the same year they are transplanted. The nurseryman reckons the age of his stock from the time of transplanting, and the age of the marketable tree from the time when the buds or grafts begin to grow. Stocks are sometimes "dressed" or trimmed before being set into the nursery. This operation consists in cutting off a fourth or third of the top, and the tap root. This causes the roots to spread and induces a vigorous growth of top, because it reduces the number of shoots; and such stocks are more expeditiously handled than long and untrimmed ones. A Manetti rose stock, dressed and ready for planting, is shown in Fig. 86. This stock was grown in France, and upon being received in this country was trimmed as it is now seen. It will now (in the spring) be set in the nursery row, and it will be budded near the surface of the ground in the summer.

86. Dressed rose stock (x¼).

Stocks should be at least three-eighths inch in diameter to be budded with ease. Just before the buds are set, the leaves are removed from the base of the stock, so that they will not interfere with the operation. They are usually rubbed off with the hand for a space of five or six inches above the ground. They should not be removed more than two or three days in advance of budding, else the growth of the parts will be checked and the bark will "set." Any branches, too, as in the quince, which might impede the work of the budder, are to be cut off at the same time. The bud is inserted an inch or two above the surface of the ground, or as low down as the budder can work. The advantage of setting the bud low is to bring the resulting crook or union where it will not be seen, and to enable it to be set below the surface of the ground when the tree is transplanted, if the planter so desires. It is a common and good practice, also, to place the bud upon the north side of the stock to shield it from the sun. A greater number of the buds will grow when set upon the north side.

The buds are taken from strong and well hardened shoots of the season's growth and of the desired variety. Usually the whole of the present growth is cut, the leaves are removed, but a part of the petiole or stalk of each leaf is left (as in Figs. 85 and 87) to serve as a handle to the bud. This trimmed shoot is then called a "stick." A stick may bear two dozen good buds when the growth has been strong, but only ten or twelve buds are commonly secured. The upper buds, which are usually not fully grown, and which are borne on soft wood, are usually discarded.

87. Stick of buds (x⅓).

The buds are cut with a thin-bladed sharp knife. Various styles of budding knives are in use, and the budder usually has decided preferences for some particular pattern. The essentials of a good budding knife are these: the very best steel, a thin blade which has a curved or half-circular cutting end, which is light, and handy in shape. The curved end of the blade is used for making the incisions in the stock. The handle of the budding-knife usually runs into a thin bone scalpel at the end, and this portion is designed for the lifting or loosening of the bark on the stock. The operation of raising the bark by means of this scalpel is often called "boning." Some budders, however, raise the bark with the blade. A good form of blade, but one seldom made, has a rounded end, the upper side

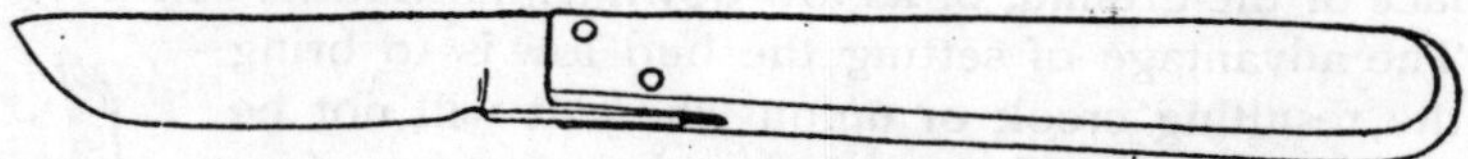

88. Budding-knife with stationary blade (x½).

of the curve being ground simply to a thin edge. This blade may be used both for cutting the bark and loosening it, thus overcoming the necessity of reversing the knife every time a bud is set. If this form of blade were commonly known it would undoubtedly soon come into favor. The blade of a common budding-knife can be ground to this shape. In the large fruit-tree nurseries of New York state, the knife shown in Fig. 88 is in common use. This is a cheap knife (costing fifteen cents or less by the dozen), with a stationary blade. When using this knife, the operator loosens the bark with the rounded edge of the blade.

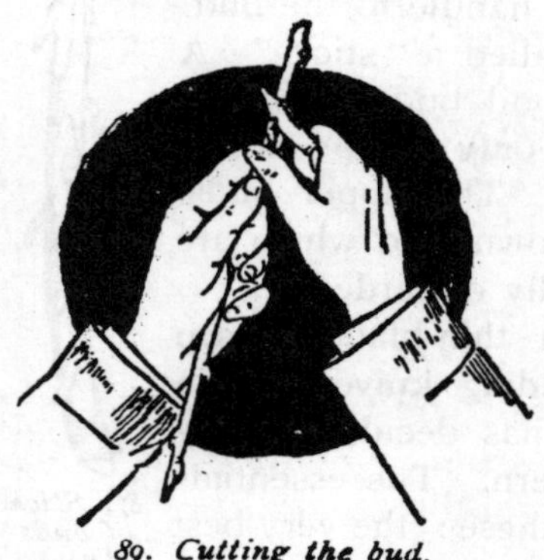

89. Cutting the bud.

The bud is usually cut about an inch long. Most budders cut from below upwards (as seen in the inverted stick in Fig. 87, and in Fig. 89), but some prefer to make a downward incision. It does not matter just how the bud is cut, if the surfaces are smooth and even, and the bud is not too thick. Some propagators cut the buds as they go, while others prefer to cut a whole stick before setting any, letting each bud hang by a bit of bark at the top, and which is cut off squarely when wanted, as is shown in Fig. 87. On a stick a fourth or three-eighths inch through the cut, at its deepest point just under the bud, is about one-fourth the diameter of the twig. A bit of wood is, therefore, removed with the bud, as shown in Fig. 85. There is some discussion as to whether this wood should be left upon the bud, but no definite experiments have been made to show that it is injurious to the resulting tree. Some budders remove the wood with the point of the knife or by a deft twist as the bud is taken from the stick. But buds appear to live equally well with wood attached or removed. The bit of wood probably serves a useful purpose in retaining moisture in the bud, but it at the same time interposes a foreign body between the healing surfaces, for the bark of the bud unites directly with the surface of the stock. Probably the very youngest portions of the wood in the bud unite with the stock, but if the budding-knife cuts deep, the denser part of the wood should be removed from the bud. This remark is particularly true, also, of all buds which are likely to be cut into the pith, as in the nut trees.

90. Preparing the stock (x½)

The wound or matrix which is to receive the bud is made by two incisions, one vertical and one transverse (Fig. 90). These are light cuts, extending only through the bark. The vertical slit is usually made first and by the rounded end of the blade. This is an inch or inch and a half long. The transverse cut is made across the top of the vertical cut by one rocking motion of the blade. The corners of the bark may be lifted a little by an outward motion of the blade so as to allow the bud to be pushed in, but unless the bark slips very freely it will have to be loosened by the end of the blade or by the scalpel on the reverse end of the handle, as previously explained. The bud is now inserted in the cleft of the bark. It is pushed down part way by the fingers, as in Fig. 91, but it is usually driven home by pushing down upon the leaf-stalk handle with the back of the knife-blade. The entire bud should pass into the cleft; or if a portion of it should project above, it should be cut off. If the bark peels freely, the bud will slip in easily and will follow the cleft, but if it sticks somewhat, more care is necessary to prevent the bud from running out. If the bark is very tight, it may have to be loosened with the knife throughout the length of the cleft; but budding should be performed, if possible, when such pains is not necessary.

91. Bud entering matrix (x½).

The bud must now be tied. The whole matrix should be closed and bound securely, as represented in Fig. 92. The string is usually started below the bud, usually being wrapped twice below the bud and about thrice above it, in fruit-trees, the lower end being held by lapping the second course over it, and the upper end being secured by drawing a bow through under the upper course, or sometimes by tying an ordinary hard knot. Care should be taken not to bind the string over the bud itself. The strings are previously cut the required length—about a foot—and the tying is performed very quickly. Any soft cord may be employed. Yarn and carpet warp are sometimes used. The most common material, at least until the last few years, has been bass-bark. This is the inner bark of the bass-wood or linden. The bark is stripped in early summer, and the inner portion is macerated or

92. The bud tied (x½).

"rotted" in water for four or five weeks. It is then removed, cut into the desired lengths, and stripped into narrow bands—one-fourth to one-half inch wide—when it may be sorted and stored away for future use. If it is stiff and harsh when it comes from the maceration, it should be pounded lightly or rubbed through the hands until it becomes soft and pliable. The best tying material which we now have is undoubtedly raffia. It is an imported article, coming from the eastern tropics (the product of the palm *Raphia Ruffia*), but it is so cheap that it is superseding even bass-bark. It is strong and pliable, and is an excellent material for tying up plants in the greenhouse, or small ones outdoors. The greatest disadvantage in its use in the budding field is its habit of rolling when it becomes dry, but it may be dipped in water a few minutes before it is taken into the field, or, better still, it may be allowed to lie on the fresh ground during the previous night, during which time it will absorb sufficient moisture to become pliable.

In two or three weeks after the bud is set, it will have "stuck" or united to the stock. The bandage must then be removed or cut. It is the common practice to draw a budding-knife over the strings, on the side opposite the bud, completely severing them and allowing them to fall off as they will. If the strings are left on too long they will constrict the stem and often kill the bud, and they also have a tendency to cause the bud to "break" or to begin to grow. The bud should remain perfectly dormant until spring, for if it should begin to grow it will be injured and perhaps killed by the winter. It should remain green and fresh; if it shrivels and becomes brown, even though it still adheres to the stock, it is worthless. Advantage can be taken, when cutting the tyings, to rebud any stocks which have failed. If the bud should begin to grow, because of a warm and wet fall or other reasons, there is little remedy except perhaps to head the shoot back if it should become long enough. If the stocks are protected by snow during winter, some of the buds at the base of the shoot may pass the cold in safety. A dormant bud, as it appears in the winter following the budding, is shown in Fig. 93. This bud was inserted in August, 1895; the picture was made in March, 1896; the bud should have started to grow in May, 1896.

93. *Dormant bud of plum* (x1).

The spring following the budding, the stock should be cut off just above the bud, in order to throw the entire force of the plant into the bud. The stock is generally, and preferably, cut off twice. The first cutting leaves the stub 4 or 5 inches long above the bud. This cutting is made as soon as the stocks begin to show any signs of activity. Two weeks later, or when the bud has begun to grow (the shoot having reached the length of an inch or two), the stock is again cut off a half-inch above the bud (Fig. 94). A greater proportion of buds will usually grow if this double heading-in is done, in outdoor conditions, than if the stock is cut back to the bud at the first operation. If the root is strong and the soil good, the bud will grow 2 to 6 feet the first year, depending much upon the species. All sprouts should be kept rubbed off the stock, and the bud should be trained to a single stem. In some weak and crooked growers, the new shoot must be tied, and some propagators in such cases cut off the stock 5 or 6 inches above the bud and let it serve as a stake to which to tie; but this operation is too expensive to be employed on common fruit trees. The stock, of course, must not be allowed to grow. Late in the season

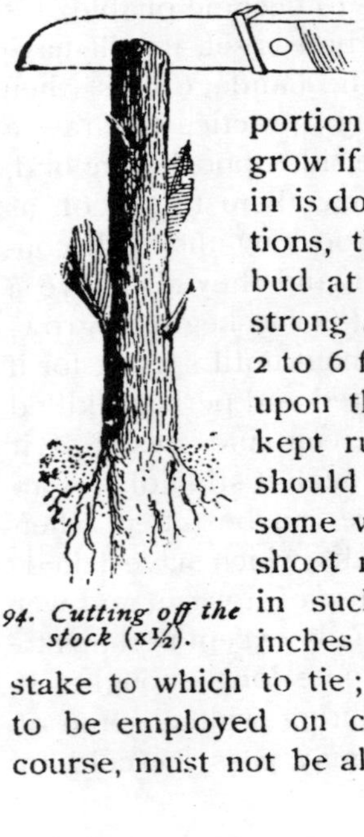

94. *Cutting off the stock* (x½).

the stock is cut down close to the bud. Peaches and some other fruits are sold after having made one season's growth from the bud, but pears, apples, and most other trees are not often sold until the second or third year.

"June budding" is a term applied to the budding of stocks in early summer, while they are yet growing rapidly. It is employed at the south, where the stocks can be grown to sufficient size by the last of June or first of July. Small stocks are usually employed—those ranging from one-fourth to one-third inch being preferred. A few strong leaves should be left on the stock below the bud, and after the bud has "stuck," the whole top should not be cut off at once, else the growing plant will receive a too severe check. It is best to bend the top over to check its growth, or to remove the leaves gradually. The bandages should not be left on longer than six to ten days if the stock is growing rapidly. To prevent the constriction of the stem, muslin bands are sometimes used instead of bass or raffia. In hot and dry climates the buds should be set an inch or two higher in June budding than in the ordinary practice, to escape the great heat of the soil. June budding is used upon the peach more than any other tree, although it can be employed for any species which will give large enough stocks from seed by the June following the sowing. In peaches, the bud will produce a shoot from 3 to 5 feet high the same season the buds are set, so that marketable budded trees can be produced in one season from the seed.

A different kind of early summer budding is sometimes performed upon apples and other fruit-trees. In this case, the stocks are one or two years old from the transplanting, the same as for common budding, but dormant buds are used. These buds are cut the previous fall or winter in the same manner as cions, and when spring approaches they are put on ice—in sawdust, sand or moss—and kept until the stocks are large enough to receive them. The particular advantage of this method is the distributing of the labor of budding over a longer season, thereby avoiding the rush which often occurs at the regular budding time. It is also a very useful means of top-working trees, for the buds start the same season in which the buds are set, and a whole season is thereby saved as compared with the common summer or fall budding.

Budders usually carry a number of "sticks" with them when they enter the nursery. These may be carried in the pocket, or thrust into the boot-leg; or some budders carry four or five sticks in the hand. The budder follows a row throughout its length, passing over those trees which are too small to work. It is a common practice to rest upon one knee while budding, as shown in Fig. 95, but some prefer to use a low stool. It is a common practice, in some nursery regions, for budders to use a low box with half of the top covered to serve as a seat, and the box is used for carrying buds, string, knives and whetstone. The tying is usually done by a boy, who should follow close behind the budder, in order that the buds shall not dry out. An expert budder will set from 1,000 to 3,000 buds a day, in good stock, and with a boy (or two of them for the latter speed) to tie. Peach stocks are more rapidly budded than most others, as the bark is firm and slips easily, and some remarkable records are made by skillful workmen.

95. Budder at work.

Budding is sometimes employed the same as top-grafting for changing over the top of an old tree from one variety to another. The buds cannot be easily inserted in very old and stiff bark, but in all smooth and fresh bark they work readily, even if the limb is three or four years old; but the younger the limb, the greater the proportion of buds which may be expected to live. Sometimes old trees are severely pruned the year before the budding is to be done, in order to obtain young shoots in which to set the buds. In fruit trees six or seven years old or less, budding is fully as advantageous as grafting. New varieties are also budded into old branches in order to hasten bearing of the bud, for the purpose of testing the variety. Here budding has a distinct advantage over grafting, as it uses fewer buds, and the wood of new sorts is often scarce.

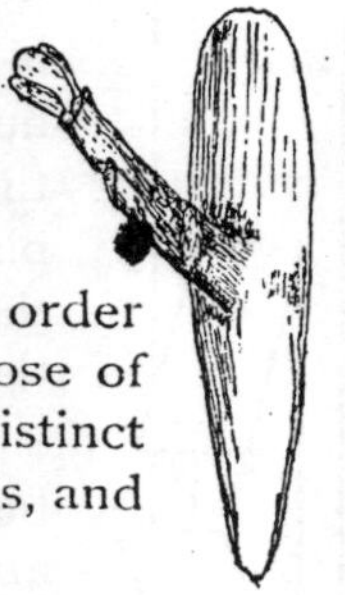

96. Prong-bud (x1).

Prong-budding.—A modification of the common shield-bud is the use of a short prong or spur in the place of a simple bud. The bud is cut in essentially the same manner as the shield-bud (Fig. 96). This is chiefly used upon the Pacific coast for nut trees, particularly for the walnut, and when the trees are dormant. The method is very much like grafting, for the stock is cut off just above the bud when the operation is performed, and the wound, in addition to being tied, is covered over with grafting wax. In budding the walnut, it is essential that nearly all the wood be removed from the bud, in order to bring as much as possible of the bark in direct contact with the stock. This is sometimes called twig-budding.

97. Plate-budding (x½).

Plate-budding is a method sometimes employed with the olive, and is probably adapted to other species. A rectangular incision is made through the bark of the stock, and the flap of bark is turned down (Fig. 97). A bud is cut of similar shape, with no wood attached, and it is inserted in the rectangular space, and is then covered with the flap, which is brought up and tied. The subsequent treatment of the bud is similar to that of the ordinary shield-bud.

A method of winter budding used at the Texas Experiment Station (Bull. 37, p. 713; Sixth Rep., 414) is evidently a modification of this plate-budding: "The method is simply to cut a slice of bark down the stock, leaving it still attached to the stock at the lower end, to help hold the bud. Part of the loose strip is then cut off and the bud fitted over the cut place with the lower end being held firmly by the part of the slip left. A piece of raffia is then tied around the bud to hold it firmly."

H-budding (Fig. 98) is a modification of plate-budding. In this method, a flap is formed both above and below, covering the bud from both ends, and allowing of more perfect fitting of the bark about the bulge of the bud.

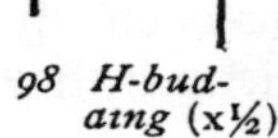

98 *H-budding* (x½).

Flute-budding. — An occasional method of budding is that known under the general name of flute-budding. In this method the bud is not covered by the bark of the stock, as in the other methods here described. Fig. 99 illustrates it. A portion of bark is removed entirely from the stock, and a similar piece is fitted into its place. When the wound extends only part way about the stem, as in the illustration, the operation is sometimes known as veneer-budding. When it extends entirely round the stem it is called ring or annular-budding. Flute-budding is usually performed late in the spring. It is best adapted to plants with very thick and heavy bark. The bud is tied and afterwards treated in essentially the same manner as in shield-budding. A species of flute-budding in which a ring of bark is slipped down upon the tip of a shoot, which has been girdled for the purpose, is called whistle- or tubular-budding.

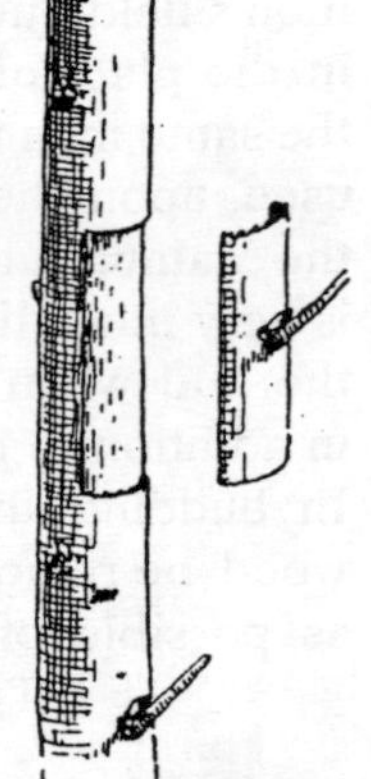

99 *Flute-budding* (x½).

100. *Chip budding* (x⅓).

Chip-budding (Fig. 100) is a method which inserts a chip of bark and wood into a mortise in the stock. It is used in spring, when the stock is dormant and the bark does not slip. It is held in place by tying, and it is better for being covered with wax.

3. GRAFTING.

Grafting is the operation of inserting a cion—or a twig comprising one or more buds—into the stock, usually into an incision made in the wood. It is divided or classified in various ways, but chiefly with reference to the position of the union upon the plant, and to the method in which the cion and stock are joined. In reference to position, there are four general classes: 1. Root-grafting, in which the stock is entirely a root. 2. Crown-grafting, which is performed upon the crown or collar of the plant just at the surface of the ground, an operation which is often confounded with root-grafting. 3. Stem-grafting, in which the cion is set on the trunk or body of the tree below the limbs, a method occasionally employed with young trees. 4. Top-grafting, or grafting in the branches of the tree. Any method of inserting the cion may be employed in these classes. The best classification, particularly for purposes of description, is that which considers methods of making the union. Some of these kinds of grafting are catalogued on pages 80 and 81. The most important methods of grafting are now to be considered; but almost endless modifications may be made in the details of the operations. The union of the cion with the stock, like the union of the bud and the stock, depends upon the growing together of the cambial tissue of the two. It is, therefore, essential that the tissue lying between the outer bark and the wood in the cion should come closely in contact with the similar tissue of the stock.

Cions are cut in fall or winter, or any time before the buds swell in spring. Only the previous year's growth is used in all ordinary cases, but in maples and some other trees, older wood may be used. In the grafting of peaches—which is very rarely done—the best cions are supposed to be those which bear a small portion of two-year-old wood at the lower end. This portion of old wood probably serves no other purpose than a mechanical one, as the recent wood is soft and pithy. It is a common opinion that cions are worthless if cut during freezing weather, but this is unfounded. The cions are stored in sand, moss or sawdust in a cool cellar, or they may be buried in a sandy place. Or sometimes, when a few are wanted for top-grafting, they are thrust into the ground beside the tree into which they are to be set the following spring. If the cions are likely to start before the spring grafting can be done, they may be placed in an ice house. Only well-formed and mature buds should be used. Sometimes flower-buds are inserted for the purpose of fruiting a new or rare variety the following year, but unless particular pains is taken to nurse such a cion, it is apt to give only very indifferent results.

101. Cion of whip-graft (x1.)

Whip-grafting.—Whip or tongue-grafting is employed only on small stocks, usually upon those one or two years old. Both the cion and stock are cut across diagonally, the cut surface extending from 1 to 2 inches, according to the size of the part. A vertical cleft is then made in both, and the two are joined by shoving the tongue of the cion into the cleft of the stock. The operation can be understood by reference to Figs. 101, 102 and 103. Fig. 101 shows the end of a cion, cut natural size. The stock is cut in the same manner, and the two are joined in Figs. 102 and 103. The parts are held firmly by a bandage—as bass bark or raffia—passed five or six times around them. If the graft is to stand above ground, the wound must be protected by

102. Whip-graft in position. (x½).

applying wax over the bandage. (Recipes for wax may be found at the end of this chapter.)

Root-grafting, especially of fruit stocks, is performed almost entirely by the whip-graft. This operation is performed in winter. The stocks, either one or two years old, are dug and stored in the fall. In January or February the grafting is begun. In true root-grafting, only pieces of roots are used, but some prefer to use the whole root and graft at the crown. In piece-root-grafting, from two to four trees are made from a single root. A piece of root from two to four inches long is used, as shown in Fig. 103. The parts are usually held by winding with waxed string or waxed bands. The string should be strong enough to hold the parts securely and yet weak enough to be broken without hurting the hands. No. 18 knitting cotton answers this purpose admirably. It should be bought in balls, which are allowed to stand for a few minutes in melted wax. The wax soon saturates the ball. The ball is then removed and laid away to dry, when it is ready for use. This waxed string will remain almost indefinitely in condition for use. Waxed bands, which are sometimes used, are made by spreading melted wax over thin muslin, which is cut into narrow strips when dry. The string is the more useful for rapid work. The grafts are packed away in sand, moss or sawdust in a cool cellar until spring, when the two parts will be firmly callused together. Some propagators are now discarding all tying of root-grafts. The grafts are packed away snugly, and if the storage cellar is cool—not above 40°—they will knit together so that they can be planted without danger of breaking apart. If the cellar is warm, the grafts will start into growth and be lost. It is very important that the cellar in which root-grafts are stored shall not become close or warm, else the grafts will

103. Root-graft.

heat or rot. Some of the characteristics of root-grafted trees are discussed in the last part of this chapter.

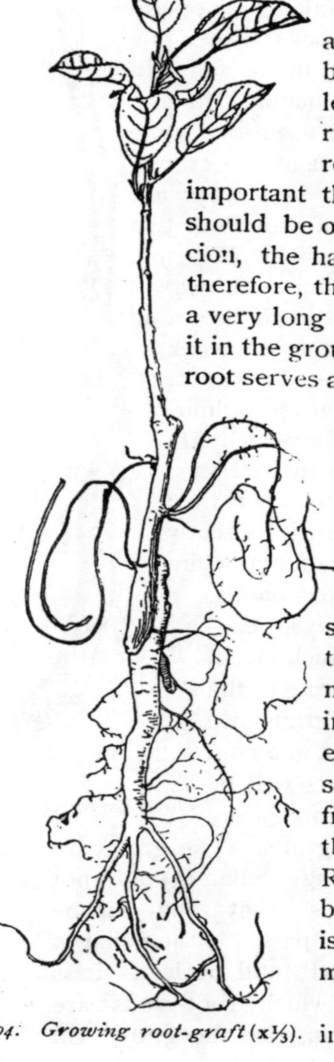

ɔ4. *Growing root-graft* (x⅓).

In common root-grafting in the east and south, the cion bears about three buds, and the root is about the same length, or perhaps shorter. The variable and unknown character of these roots as, regards hardiness, renders it important that, in very severe climates, roots should be obtained from the same plant as the cion, the hardiness of which is known. It is, therefore, the practice in the prairie states to use a very long cion—8 inches to a foot—and to set it in the ground up to the top bud. The piece of root serves as a temporary support, and roots are emitted along the cion. When the tree is ready for sale the old piece of root is often removed, or sometimes it falls away of itself. In this manner own-rooted trees are obtained, and it is for this reason that root-grafting is more universally practiced west of the Great Lakes than budding is. Even cions of ordinary length often emit roots, as seen in Fig. 104, but such cions are not long enough to reach into uniformly moist soil. In practice, some varieties of fruit trees are found to emit roots from the cion more readily than others. Root-grafting is often cheaper than budding, as it is performed when labor is cheap, and two or more trees are made from one stock.

Cuttings may be used as stocks in those instances in which a variety which grows readily from cuttings unites quickly with a variety which does not grow from cuttings. Fig. 105 illustrates such a case. The stock, or cutting, is the true Downing mulberry, which strikes root readily. The cion is any of the varieties of *Morus alba* or *M. rubra*, like the New American or Hicks, which roots with difficulty from cuttings. In this instance, the buds have been cut from the stock to prevent it from suckering.

Any sharp and strong thin-bladed knife may be used for the making of whip-grafts. For small and tender plants, a common budding-knife is sufficient, but it is too light for most work. A favorite style of knife for root-grafting is shown in Fig. 106. It is much like a shoe-knife, with large, cylindrical handle and a stationary blade. These knives can be had by the dozen for about twenty-five cents apiece.

Modified Whip-grafts.—There are many modifications of the whip-graft. One of them (Fig. 107), used for the grape, is described by Lodeman in "The Grafting of Grapes" (Bulletin 77, Cornell Experiment Station): "Fig. 107 represents a form of grafting which is quite common in Italy. The stock is cut off at an angle an inch or two below the surface of the soil, and is then split downward, beginning a little above the center of the cut surface. This downward cut is made at a slight angle to the grain, in order to prevent splitting. In true tongue or whip-grafting the cion is prepared in the same manner as the stock; but in the graft shown in the figure, a portion of the bark is

105. Whip-grafted cutting (x⅓).

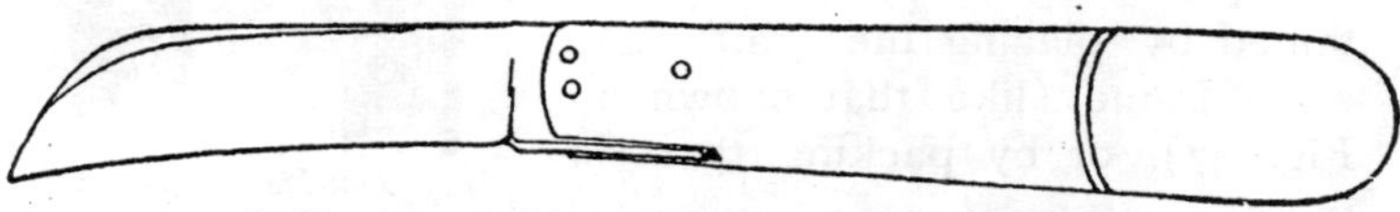

106. Grafting-knife with stationary blade (x½).

first removed, and from the lower end of this cut another is made inward and upward, in order to form the tongue. The cion is not cut in two when the tongue is made, as is the stock, but it extends below and also takes root. Cion and stock are then united, as shown in Fig. 107, care being take to have the cambium layers in contact on one side. When cuttings or parts of equal diameters are grafted by the tongue-graft, the layers on both sides may be placed together. The tying of grafts is advisable when small wood is used, but large stocks, when cut below the ground, scarcely require this precaution. When the operation is finished, the soil is heaped up, as in cleft-grafting."

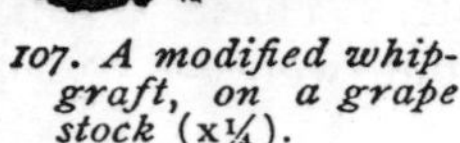

107. A modified whip-graft, on a grape stock (x¼).

An old-fashioned modification of the whip-graft leaves the end of the cion 4 or 5 inches long, so that it may project downwards into a bottle or dish of water, thereby absorbing sufficient moisture to maintain the cion until it unites with the stock. Another modification, with the same purpose in view, is to allow the ends of the tying material to fall into the water. These methods are called "bottle-grafting" in the books. They are really of no account, although they might be employed for certain difficult subjects amongst ornamental plants; but even there, better results can be obtained by placing the grafts in a close frame (like that shown in Fig. 47), or by packing them in moss.

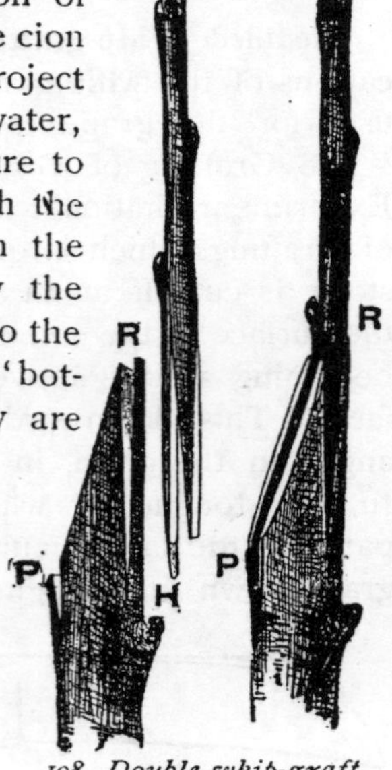

108. Double whip-graft (x⅔).

A "double whip-graft" is shown in Fig. 108. In this method, the cion is cut upon one side into a wedge, and upon the other with a long tongue (H). The stock is provided with two clefts, at R and P. This cion, having two supports in the stock, forms a most intimate contact with its host; but it is too slow, and the rewards too slight, to warrant its general use. This is sometimes, but erroneously, called a saddle-graft.

Saddle-grafting.—Saddle-grafting is a simple and useful method for the shoots of small growing plants. The stock is cut to a wedge-shape end by two cuts, and the cion is split and set upon the wedge (Fig. 109). The union is then tied and waxed in the same way as exposed whip-grafts. It is oftenest employed when a terminal bud is used, as the wood in such cions is usually too weak to work easily with a tongue.

109. Saddle graft (x½).

Splice-grafting.—The simplest form of grafting is that shown in Fig. 110, in which the two parts are simply cut across diagonally and laid together. The parts are held only by the string, which, together with the wax, is applied in the same way as upon the whip-graft. Splice-grafting is frequently used upon soft or tender wood which will not admit of splitting. It is adapted only to small shoots.

110. Splice-grafting (x½).

Veneer-grafting.—Fig. 111 shows a style of grafting which is much used, particularly for ornamentals and for rare stocks which are grown in pots. An incision is made upon the stock just through the bark and about an inch long (*A*, Fig. 111), the bit of bark being removed by means of a downward sloping cut at its base. The base of the cion is cut off obliquely, and upon the longest side a portion of bark is removed, corresponding to the portion taken from the stock. The little tongue of bark on the stock covers the base of the cion when it is set. The cion is tied tightly to the stock (*B*, Fig. 111), usually with raffia. This method of grafting makes no incision into the wood, and all wounded surfaces are completely covered by the matching of the cion and stock. (See Fig. 83, page 88, and compare it with the picture of a whip-graft union in Fig. 84, page 89.) It is not necessary, therefore, to wax over the wounds, as a rule. If used in the open, however, wax should be used. The parts grow together uniformly and quickly, making a solid and perfect union, as shown at *D*. So far as the union of the parts is concerned, this is probably the ideal method of grafting. This method, which is nothing but the side-graft of the English gardeners with the most important addition of a longer tongue on the stock, is known by various names, but it is oftenest called veneer-grafting in this country.

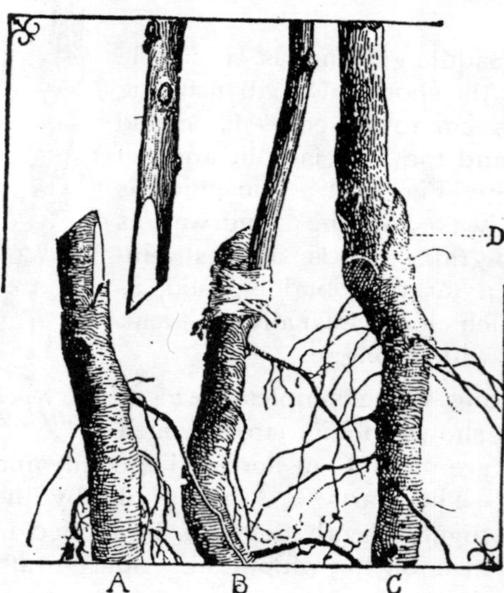

111. *Veneer-grafting* (x½).

Veneer-grafting is employed mostly from November to March, upon potted plants. Stocks which are grown outdoors are potted in the early fall and carried over in a cool house or pit. The cion is applied an inch or two above the surface of the soil, and the stock need not be headed back until the cion has united. (See Fig. 112.) Both dormant and growing cions are used. All plants in full sap must be placed under a frame in the house, in which they can be almost entirely buried with sphagnum, not too wet, and the house must be kept cool and rather moist until the cions are well established. Some species can be transferred to the open border or to nursery rows in the spring, but most plants which are grafted in this way are handled in pots during the following season. Rhododendrons, Japanese maples and many conifers are some of the plants which are multiplied by veneer-grafting. Such plants are usually laid upon their sides in frames (Fig. 47) and covered with moss for several days, or until healing begins to take place. This method, when used with hardy or tender plants, gives a great advantage in much experimental work, because the stock is no at all injured by a failure, and can be used over again many times, perhaps even in the same season; and the manipulation is simple, and easily acquired by inexperienced hands.

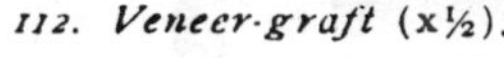

112. Veneer-graft (x½).

Side-grafting.—There are various methods of inserting a cion into the side of a stock without cutting off the stock. One of the best styles is shown in Fig. 113. The example upon the right shows the cion set into an oblique cut in the stock, and that upon the left shows the lower part of a thin-bladed chisel, with a bent shank, used for

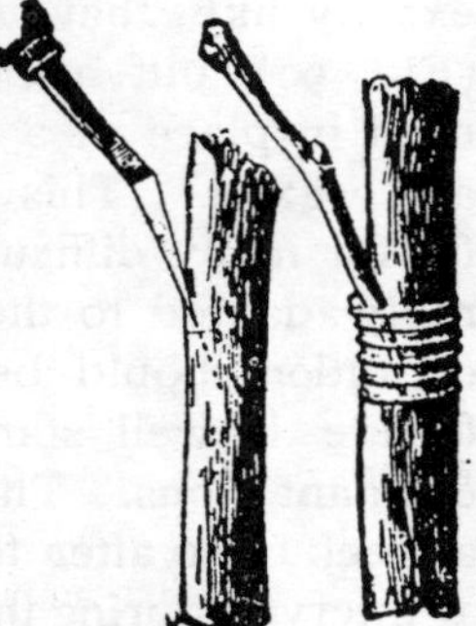

113. Side-graft (x½).

making the incision. An ordinary chisel or a knife may be used, however. The incision should be about an inch deep. The cion is cut wedge-shape, as for cleft-grafting, and it is pressed into the incision until its cut surfaces are concealed in the stock. The wound is then tied, and, if it is above ground in the open, it is waxed. The stock is headed back vigorously to aid in deflecting a part of the energy into the cion. This method of grafting may be used to good advantage upon rather small grape stocks, below the surface of the ground.

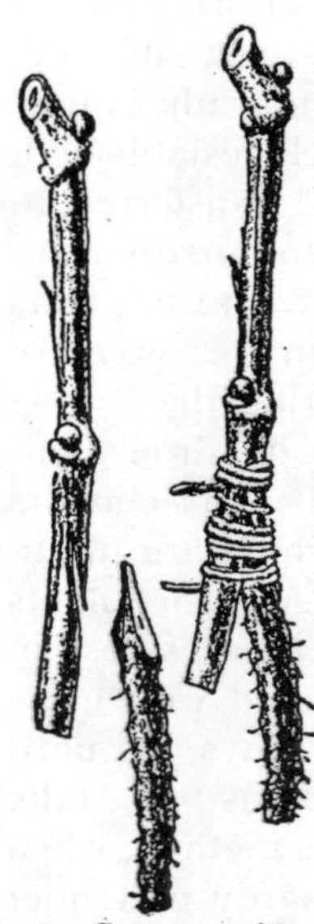

114. Cutting side-graft (x⅓).

A modification of this style of side-grafting is the "cutting side-graft," shown in Fig. 114. This is adapted to root-grafting, particularly of the grape. The stock is cut wedge-shape, and is inserted into an oblique incision in the cion.

A side-graft which is a combination of budding and grafting is shown in Fig. 115. The incision in the stock is exactly like that made for shield-budding (Fig. 90), but a cion, cut wedge-shape, is used in place of a bud. The graft is tied and waxed. This style of grafting is useful for many difficult subjects. It is admirably adapted to the mulberry, in which the operation should be performed just as the foliage is well started in the spring, with dormant cions. The stock is headed back a week or so after the cion is set, and again at intervals during the season. The cion will often make sufficient growth the first season to form a salable tree by fall. Purple and weeping beeches may be grafted in this

115. Shield-grafting, or cion-budding (x1).

same fashion, except that the operation should be done in late summer or fall, with freshly cut cions, much the same as summer budding is done.

Inlaying.—There are various styles of grafting in which a piece of wood is removed from the stock and a cion is cut to fill the cavity. The following methods described by Lodeman for the grafting of grapes (Bulletin 77, Cornell Experiment Station), will serve as a type of the class: "The stock is cut off, as for cleft grafting. In place of splitting the stub, one or two V-shaped grooves are made in it (Fig. 116). These grooves are made by means of an instrument especially designed for the purpose. It is shown in Fig. 117. The tip cuts out the triangular part. In the blade itself is a part which is bent at the same angles as the parts forming the tip. This indented portion of the blade is used for cutting away the end of the cion, and with very little practice an almost perfect fit of the two parts can be made. The one or two cions are then placed upon the stock and are firmly tied there. The tying material should be of such a nature that it will decay before there is any danger of strangling the cions. Raphia does very well, as does also bast. No. 18 knitting cotton, soaked in boiling grafting wax, may be used with entire satisfaction. The ligatures should be made as tight as possible. Although this method of grafting is not so commonly used as others, it still possesses some decided advantages for grape vines. It is a much simpler and more satisfactory method than cleft-grafting in very curly wood. The tying is a slow process, and for straight-grained wood the cleft graft is to be preferred. It is also open to the objection of requiring the shoots to be staked or tied to some support, for the wind is apt to break the point of union more easily than with

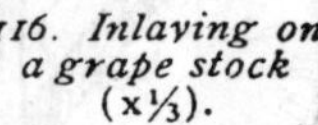

116. Inlaying on a grape stock (x⅓).

117. Inlaying tool (x¼).

other methods. A good union admits of a very strong growth, and if the above precautions are kept in mind the vines will equal those produced by the more common methods."

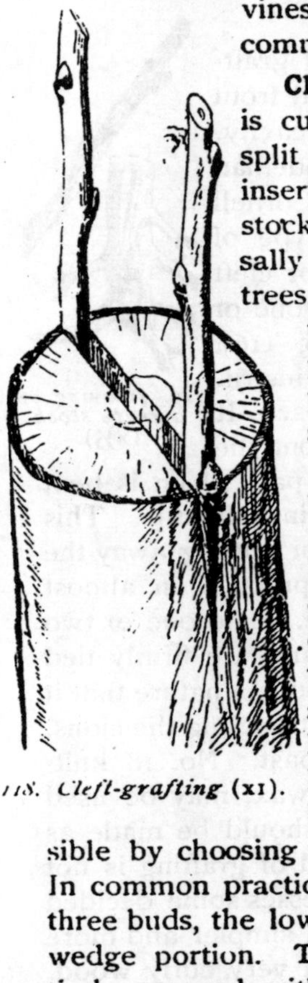

Cleft-grafting.—In cleft-grafting, the stock is cut off squarely and split, and into the split a cion with a wedge-shape base is inserted. It is particularly adapted to large stocks, and is the method almost universally employed for top-grafting old trees, its only competitor being the bark-graft described on page 129. Fig. 118 illustrates the operation. The end of the stock, technically called a "stub," is usually large enough to accommodate two cions, one upon either side. In fact, it is better to use two cions, not only because they double the chances of success, but because they hasten the healing of the stub. Cleft-grafting is at best a harsh process, especially upon large limbs, and its evils should be mitigated as much as possible by choosing small limbs for the operation. In common practice, the cion (Fig. 119) contains three buds, the lowest one standing just above the wedge portion. This lowest bud is usually entirely covered with wax, but it pushes through without difficulty. In fact, being nearest the source of food and most protected, its chances of living are greater than those of the higher buds. The sides of the cion must be cut smooth and even. A single draw cut on each side with a sharp blade is much better than two or three partial cuts. A good

118. Cleft-grafting (XI).

119. Cleft-graft cion (XI).

grafter makes a cion by three strokes of the knife, one to cut off the cion and two to shape it. The outer edge of the wedge should be a little thicker than the inner one, so that the stock will bind upon it and hold it firm at the point where the union first takes place.

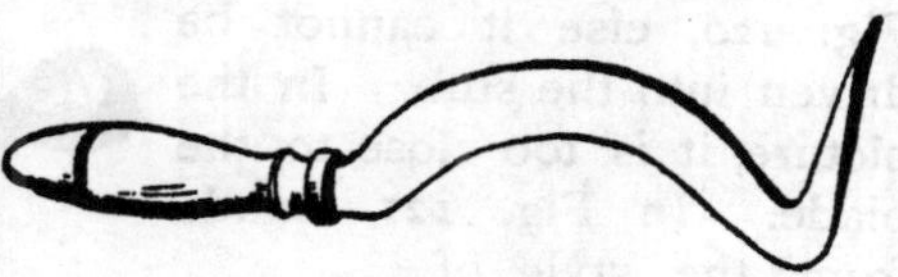

120. Cleft-grafting knife (x1-5).

These cions are taken in late fall or winter, or very early spring, and are kept in the same manner as directed on pages 107 and 108.

The stock or stub must be cut off square and smooth with a sharp and preferably fine-toothed saw. If one desires to be especially careful in the operation, the end of the stub, or at least two opposite sides of it, may be dressed off with a knife, so that the juncture between the bark and the wood may be more easily seen. Professional grafters rarely resort to this dressing, however. The stub is then split to the depth of an inch and a half r two inches. Various styles of grafting-knife are used to split the stub. The best one is that shown in Fig. 120. It is commonly made from an old file by a blacksmith. The blade is curved, so that the bark of the stub is drawn in when the knife is entering, thereby lessening the danger of loosening the bark. Another style of knife is illustrated in Fig. 121. In this tool, the cutting edge is straight, and, being thinner than the other tool, it tends rather to cut the stub than to split it. Upon the end of these knives is a wedge, about 4 or 5 inches long, for opening the cleft. The wedge is driven into the cleft and allowed to remain while the cions are being placed. If the cleft does not open wide enough to allow the cions to enter, the operator bears down on the

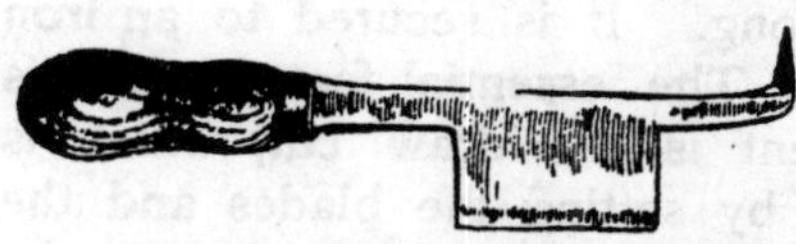

121. Cleft-grafting knife (x⅙).

handle of the knife. It is important that the wedge stand well away from the curved blade in the knife shown in Fig. 120, else it cannot be driven into the stub. In the picture, it is too close to the blade. In Fig. 121—made from the style of knife most commonly seen in the market the wedge is too short for most efficient service.

122. Hoit's grafter.

There are various devices for facilitating the operation of cleft-grafting, but none of them have become generally popular. One of the best is Hoit's device (Fig. 122), which cuts a slot into the side of the stub. The machine is held in place by a trigger or clamp working in notches on the under side of the frame. The upper handle is then thrown over to the right, forcing the knife into the stub. This is a Californian device. A very good grafting-knife for small stocks or trees in nursery row is shown in Fig. 123. This is the Thomas knife. The larger arm is made entirely of wood. At its upper end is a grooved portion, into which the blade closes. This blade can be made from the blade of a steel case-knife, and it should be about 2½ inches long. It is secured to an iron handle. The essential feature of this implement is the draw cut, which is secured by setting the blades and the pivot in just the position shown in the figure. The stock is cut off by the shears, and the cleft is then made by turning the shears up and making a

123. Thomas' grafting-knife.

vertical cut. The cleft is, therefore, cut instead of split, insuring a tight fit of the cions. This tool is particularly useful upon hard and crooked-grained stocks.

The cions must be thrust down, in the cleft, to the first bud, or even deeper, and it is imperative that they fit tight. The line of separation between the bark and wood in the cion should meet as nearly as possible the similar line in the stock. The cions are usually set a trifle obliquely, the tops projecting outwards, to insure the contact or crossing of the cambium layers. Writers usually state that it is imperative to success to have the exact lines between the bark and wood meet for at least the greater part of their length, but this is an error. The callus or connecting tissue spreads beyond its former limits when the wounds begin to heal. The most essential points are rather to be sure that the cion fits tightly throughout its whole length, and to protect the wound completely with an air-tight covering. The practice must be modified, of course, to suit the stock and the occasion. Sometimes rooted cuttings of grapes are cleft-grafted (Fig. 124), and these, being in the ground, are not waxed, and it is difficult to split the stub deep enough to allow the cion to be thrust in far. If the stub, in this case, has little elasticity after being split, it should be tightly wound to keep the cion in place. An old grape stock, cleft-grafted, and then covered with earth, is seen in Fig. 125. These covered grape stubs are usually not waxed. This is the common, and generally the best, method of grafting the grape.

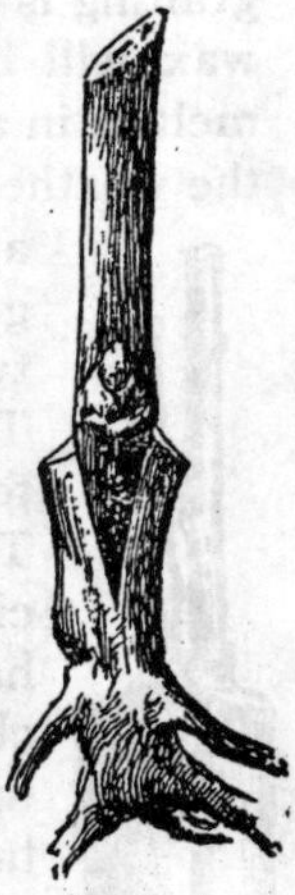

124. Rooted grape cutting cleft-grafted (x⅓).

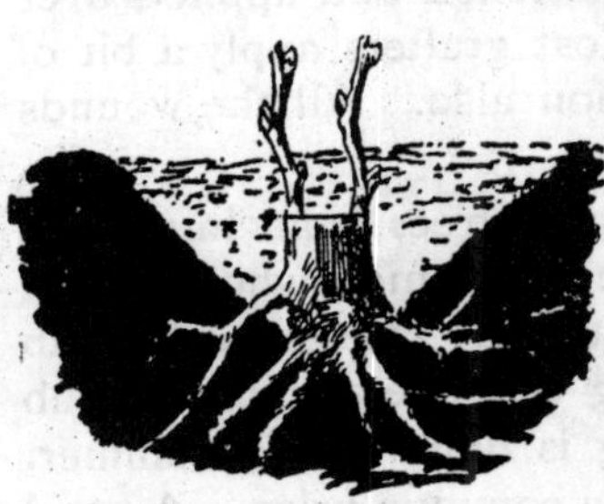

125. Cleft-grafting on old grape stock.

The wounds must now be covered with wax. Fig. 126 illustrates a stub after the covering has been applied. If the grafting is done in early spring, when the weather is cold, the wax will have to be applied with a brush. The wax is melted in a glue-pot, which is carried into the tree. But if the weather is warm enough to soften the wax, it should be applied with the hands. The hands are first greased to prevent the wax from sticking. The two side or vertical portions are applied first. The end of the mass of wax in the hand is flattened into a thin portion about a half inch wide. This portion is then laid over the lower bud of the cion and held there by the thumb of the other hand, while the wax is drawn downwards over the cleft, being pressed down firmly upon the bark by the thumb of the first hand. The wax gradually tails out until it breaks off just below the lowest point of the cleft. The flattened upper part is then wrapped around the cion upon either side, completely and tightly encircling it. A simple deft wrapping of the wax about the cion makes a tighter joint than can be secured in twice the time by any method of pinching it into place. Another portion of wax is now flattened and applied over the end of the stub. Most grafters apply a bit of wax to the tops of the cion also. All the wounds must be covered securely.

126. A waxed stub (x½).

The top-grafting of large trees is an important operation, and there are many men who make it a business. These men usually charge by the stub and warrant, the warrant meaning that one cion of the stub must be alive when the counting is done late in summer. From two to three cents a stub is a common price. A good grafter in good "setting" can graft from 400 to 800 stubs a day and wax them himself. Much depends upon the size of the trees, their shape, and the amount of pruning which must be done before the grafter can work in them handily.

Every man who owns an orchard of any extent should be able to do his own grafting. The most important factor in the top-grafting of an old tree is the shaping of the top. The old top is to be removed during three or four or five years, and a new one is to be grown in its place. If the tree is old, the original plan or shape of the top will have

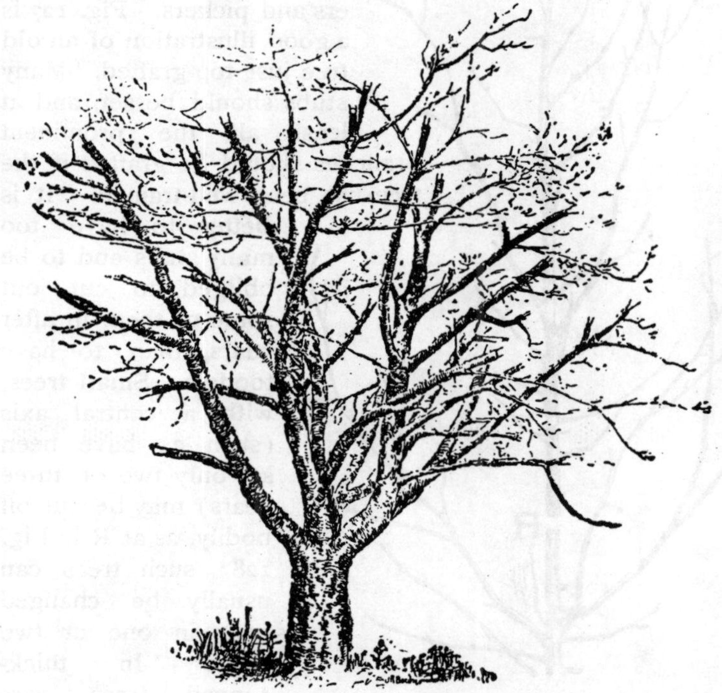

127. Top-grafted old tree.

to be followed in its general outlines. The branches should be grafted, as a rule, where they do not exceed an inch and a half in diameter, as cions do better in such branches, the wounds heal quickly, and the injury to the tree is less than when very large stubs are used. The operator should endeavor to cut all the leading stubs at approximately equal distances from the center of the tree; and then, to prevent the occurrence of long and pole-like branches, various minor side-branches should be grafted. These will serve to fill out the new top and to afford footholds for pruners and pickers. Fig. 127 is a good illustration of an old tree just top-grafted. Many stubs should be set, and at least all the prominent branches should be grafted if the tree has been well-trained. It is better to have too many stubs and to be obliged to cut out some of them in after years, than to have too few. Small trees, with a central axis (such as have been set only two or three years) may be cut off bodily, as at R in Fig. 128; such trees can usually be changed over in one or two years. In thick-topped trees, care must be exercised not to cut out so much foliage the first year that the inner branches will sunburn. All large branches which must be sacrificed ought to be cut out when the grafting is done, as they increase in diameter very rapidly after so much of the top is removed.

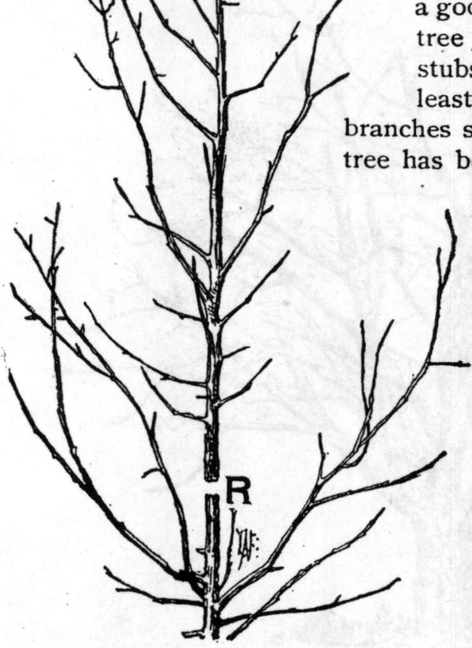

128. Stub for top-grafting a young tree.

A horizontal branch lying directly over or under another

should not be grafted, for it is the habit of grafts to grow upright rather than horizontal in the direction of the branch; and it is well to split all stubs on such branches horizontally, that one cion may not stand directly under another. The habit of growth of the cion is well shown in Fig. 129. This illustrates the form and direction of the original branch, and also the direction which the yearling grafts have taken. It is evident, therefore, that a top-grafted tree is narrower and denser in top than the tree originally was, and that careful pruning is required to keep it sufficiently open. Each graft is virtually a new tree-top placed into the tree, and for this reason, if for no other, the common practice of grafting old trees close down in the large limbs is seen to be pernicious

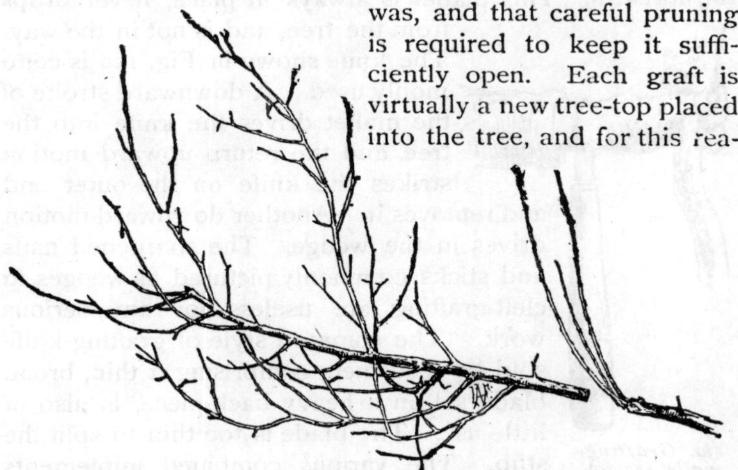

129. Showing the upright direction of a graft in a horizontal limb.

Top-grafting is performed in spring. The best time is when the leaves are pushing out, as wounds made then heal quickly, and cions are most apt to live. But when a large amount of grafting must be done, it is necessary to begin a month, or even two, before the leaves start. On the other hand, the operation can be extended until a month or more after the leaves are full-grown, but such late cions make a short growth, which is likely to perish the following winter. Professional grafters usually divide their men into three gangs,—one to do the cutting of the stubs, one to set the cions, and one to apply the wax. The cions are all whittled before the grafter enters the tree. They are then usually moistened by dipping into a pail of water, and are carried in a high side-pocket in the jacket. The handiest mallet is a simple club or billy, a foot and half long, hung over the wrist by a loose soft cord (Fig. 130). This is brought into the palm of the hand by a swinging motion of the forearm. This mallet is always in place, never drops from the tree, and is not in the way. The knife shown in Fig. 120 is commonly used. A downward stroke of the mallet drives the knife into the tree, and the return upward motion strikes the knife on the outer end and removes it. Another downward motion drives in the wedge. The sharpened nails and sticks commonly pictured as wedges in cleft-grafting are useless for any serious work. The common style of grafting-knife sold by seedsmen, comprising a thin, broad blade set in a heavy back-piece, is also of little use. The blade is too thin to split the stub. The various combined implements which have been devised to facilitate cleft-grafting are usually impracticable in commercial grafting.

130. Grafting-mallet (XI-10).

It is very important that the cleft-graft should be kept constantly sealed up until all the wounded surfaces are completely covered with the healing tissue. Old wood never heals. Its power of growth is completed. If a limb of an apple tree a half inch or more in diameter is cut off, the heart or core of the wound will be found to be incapable of healing itself. It is covered over by the callus tissue which rolls in from the cambium underneath the bark. The wound becomes hermetically sealed by the new tissue. In the meantime, the wound should be kept antiseptic by some dressing, like wax or paint, to prevent decay. In cleft-grafts, the surfaces should be covered with wax every year until they are closed in by the new tissue. In most instances, the wax will loosen during the first season, and sometimes it falls off.

The character of the healing process is well depicted in Figs. 131, 132, 133. In Fig. 131 is shown a yearling graft of apple. The strip of wax along the side of the cleft is seen to have split with the enlargement of the branch, and the cleft has filled up with tissue and is now safe from infection of disease or rot. The roll of healing tissue upon the end of the stub is seen about the border of the wound. This tissue has not yet covered up the cleft across the end of the stub, and this cleft, if exposed to the weather, is a fertile place for the starting of decay, for it does not unite except along the sides of the stub beneath the bark. When this stub is split through, following the cleft, we may readily distinguish the location of the healing tissues, Fig. 132. The ends of the cions are at E, and they are now simply inactive and nearly lifeless bits of wood. The new or healing tissue has been built up on the outward side of the cions. On the left, this deposition of new tissue may be traced as far down as H, whilst it is thick and heavy at E and above. The whole interior portion of the stub, represented by the dark shading, is dead tissue, which will soon begin a rapid process of

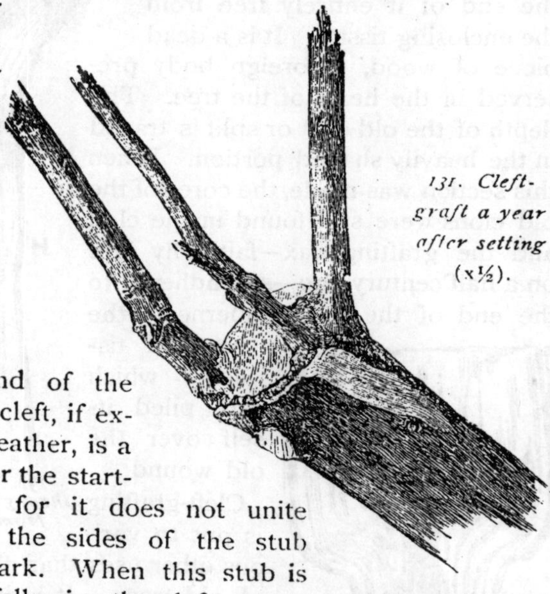

131. Cleft-graft a year after setting (x½).

decay unless it is well protected from the weather. In time, the old stub becomes hermetically sealed by the reparative tissue. Fig. 133 shows a section of an apple graft nearly fifty years old. The original stub, about an inch in diameter, is seen in the center, the end of it entirely free from the enclosing tissue. It is a dead piece of wood, a foreign body preserved in the heart of the tree. The depth of the old cleft or split is traced in the heavily shaded portion. When this section was made, the cores of the old cions were still found in the cleft and the grafting-wax—faithfully laid on a half century ago—still adhered to the end of the stub, underneath the mass of tissue which had piled itself over the old wound.

132. The stub 131 split through the cleft, and seen from the opposite side.

Cleft-grafting is put to various other uses than the top-grafting of old trees. It is in common use on soft and fleshy stocks, as cactuses, and various fleshy roots. Fig. 134 shows a cleft-graft on cactus. The cion is held in place with a pin or cactus spine, and it is then bound with raffia or other cord. Waxing is not necessary. A similar graft is often made on peony roots. The cleft in the thick

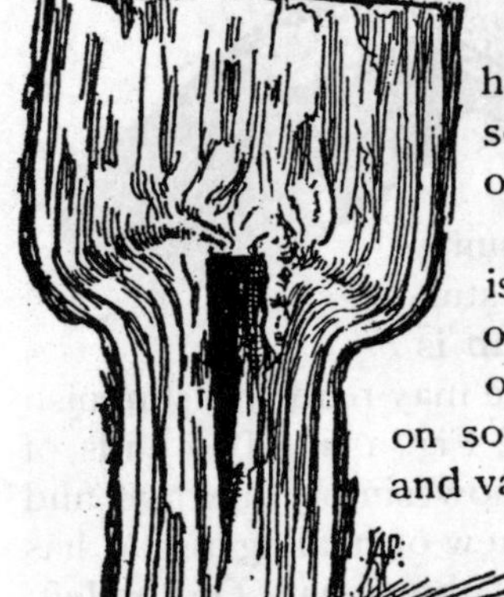

133. Section of an old cleft-graft on an apple tree.

root is cut with a knife, and the stock is bound up securely, usually with wire, as cord, unless waxed, rots off too quickly. Wax is not used, as the graft is buried to the top bud. The peony is grafted in summer. Dahlias are often grafted in the same fashion, although some operators prefer, in such fleshy subjects, to cut out a section from the side of the stock to receive the cion, rather than to make a cleft, much as in the process of inlaying illustrated in Fig. 116. Hollyhocks, ipomeas, gloxinias and other thick-rooted plants may be similarly treated.

134. Cleft-graft of cactus (x⅓).

Bark-grafting.—A style of grafting suited to large trees is shown in Fig. 135. The stock is not cleft, but the cions are pushed down between the bark and wood. The cions must be cut very thin, so that they will not break the bark on the stock. Fig. 136 represents a good style of cion. It is cut to a shoulder upon either side. Several cions can be placed in a single stub, and as no splitting is necessary, it is a useful method for very large limbs. It is especially useful in repairing trees when very large branches are broken off. The broken stub is sawn off smooth, and a dozen or more cions may be set around it. Only a few of them should be allowed to remain after the wound has been healed. Bark-grafting can be performed to advantage only when the bark peels readily. The cions should be held in place by a tight bandage, as seen in Fig. 135, and then wax should be applied in essentially the same manner as for cleft-grafting. This is sometimes called crown-grafting.

135 Bark-grafting (x⅓).

A special form of bark-grafting is sometimes employed for covering girdles about the base of an old tree, made by mice, gophers or rabbits. The edges of the bark are trimmed, and cions are cut a couple of inches longer than the width of the girdle, and they are sharpened at both ends. One end is inserted under the bark below the girdle and the other above it. The cions are placed close together entirely around the tree. The two ends are held firmly in place by tying, and the line of union is then waxed over. This operation is said to be necessary to keep up the connection between the root and the top, but this is in most cases an error, unless the girdle extends into the wood. A good dressing of wax or clay, held on with stout bandages, is usually much better than the grafting. This method of grafting is sometimes, but erroneously, called inarching. A complete bark girdle made during the spring or early summer will usually heal over readily if it is well bandaged; and in some cases even the bandage is not necessary.

136. Cion for bark-grafting (x⅔).

Herbaceous-grafting.—In the preceding pages, the discussions have had to do with cions which are dormant or at least well hardened, and with stocks which contain more or less hard woody substance. But herbaceous shoots can be grafted with ease. All such plants as geraniums, begonias, coleuses and chrysanthemums can be made to bear two or more varieties upon the same individual. Almost any style of grafting can be employed, but the veneer, cleft and saddle-grafts are preferred. Shoots should be chosen for stocks which are rather firm, or in the condition for making good cuttings. The cions should be in a similar condition, and they may be taken from the tips of branches or made of a section of a branch. The union should be bound snugly with raffia, and the plant set in a propagating-frame (Fig. 47 illustrates a good one), where it must be kept close for a few days. It is not necessary, in most cases, to use wax, and upon some tender stocks the wax is injurious. Moss may be bound about the graft, but unless the union is first thoroughly covered by the bandage, roots may start into the moss and the parts may fail to unite. The growing shoots of shrubs and trees can also be grafted, but the operation is rarely employed. In various coniferous trees (as pines and spruces) the young shoots are sometimes cleft or saddle-grafted in May, the parts being well bandaged with waxed muslin or raffia, and shaded with paper bags. The walnut and some other trees which do not work readily are sometimes treated in this manner.

A little known species of herbaceous-grafting is the joining of parts of fruits. It is easily performed upon all fleshy fruits like tomatoes, apples, squashes and cucumbers. When the fruit is half or more grown, one-half is cut away and a similar half from another fruit is applied. Better results follow if the severed side of the parent or stock fruit is hollowed out a little, so as to let the foreign piece set into the cavity. The edges of the epidermis of the stock are then tied up closely against the cion by means of bass or raffia. The two parts are securely tied together, but no wax is required. This operation succeeds best under glass, where conditions are uniform, and where winds do not move the fruits.

Even leaves may be used as stocks or cions. Any such succulent and permanent leaves as those of the house-leeks, crassula, and the like, may have young shoots worked upon them, and leaves which are used as cuttings can often be made to grow on other plants.

Seed-grafting.—A novel kind of grafting has been described in France by Pieron, which consists in using a seed as a cion. This has been used upon the grape. A seed is dropped into a gimlet-hole made near the base of the vine while the sap is rising in the spring. The seed germinates, and after a time the plantlet unites with the stock.

Cutting-grafting.—Cuttage and graftage may be combined in various ways. Cuttings of plants which root with difficulty are sometimes grafted upon those which root easily. A good example is seen in Fig. 105. When the plants are transplanted. the following autumn or spring, the nurse or stock can be removed, the cion having taken root. Root-grafting, described on a previous page (see Figs. 103, 104, 114), is virtually a grafting of cuttings. In other cases, union with an uncongenial stock is facilitated by allowing the cion to project downwards beyond the point of union, and to stand in the soil or moss or dish of water. (See, also, page 112.) Fig. 137 is a good illustration of the practice. The cion extends into the soil nearly as far as the root itself. After union has taken place, the lower part of the cion is removed. This method can be used for some magnolias, mulberries, birches, and many other plants of which some kinds root with more or less difficulty. "Bottle-grafting," described in most of the books, is essentially this method, modified by letting the end of the cion, or a portion of the bandage, drop into a bottle of water.

137. *Cutting-grafting* (x⅓).

Inarching. –Inarching, or grafting by approach, is the process of grafting contiguous plants or branches while the parts are both attached to their own roots. When the parts are united, one of them is severed from its root. Fig. 138 explains the operation. In this case, the larger plant (upon the left) is designed for the stock. When the smaller plant has united, it is cut off just below the union and it thenceforth grows upon the other plant. Limbs of contiguous trees are sometimes grafted in this way. It is the process employed by nature in what is called natural grafting (Fig. 82). Grape-vines are often inarched. A thrifty young branch of a fruit tree may be inarched into the stem of a fruit upon the same tree, thus supplying the fruit with additional food and causing it to grow larger than it might if untreated.

To join the parts, it is only necessary to remove the

barks between the stock and cion and then tie the two together snugly. The details are shown in Fig. 139. In M, a branch C, is joined at O to the stock H. Other branches, like T, might be similarly treated. In N, the method of cutting the conjoined surfaces is explained at R. If outdoors, the junction should be waxed over; and it is then necessary, also, to secure the branches in such manner that the wind cannot loosen them. The parts are sometimes joined by a tongue, after the manner of a whip-graft, but this is rarely necessary. Oranges and camellias were often propagated by inarching in the old practice, but this work is now much more easily done by the veneer-graft.

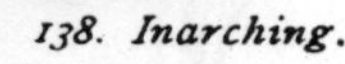

138. Inarching.

Double-working.—Grafting upon a grafted tree is known as double-grafting or double-working. It is employed for the purpose of growing a variety upon an uncongenial root, or of securing a straight and vigorous stock for a weak and poor grower. The operation may be either grafting or budding. It is more commonly the latter. Some sorts of pears do not unite well with the quince, and if it is desired to secure dwarfs of these varieties, some variety which unites readily with the quince must first be put upon it.

139. Details of inarching.

The Angouleme takes well to the quince, and upon Angouleme dwarfs the Seckel and some other varieties are often worked. In double-working dwarf pears, it is imperative that both unions be very close to the ground. The piece of interposed wood is not more than one or two inches in length.

The second cion is usually set after the first one has grown one season, although both may be set at the same time. Double-grafting for the purpose of securing a better growth is often practiced. The Canada Red apple, for instance, is such a poor grower that it is often stem-worked or top-worked upon the Northern Spy or some other strong stock. The Winter Nelis and the Josephine de Malines pears are often double-worked for the same reason. Fig. 140 shows the top of a double-worked tree. In this instance, the body of the tree is two years old and is itself a graft or bud upon a seedling root. The second variety is grafted at the point where it is desired to start the permanent top of the tree, by whip-grafting in this instance. The figure on the left shows the two-year-old top growing from this cion. The length of the cion is comprised inside the dotted lines, and this region is enlarged in the figure on the right. The base of the cion was at T—below which is stock—and the top at N. The upper scar at N is the top of the cion itself, but the other scars show where superfluous twigs were removed after the cion had grown a year. This type of double-working of fruit-trees is to be recommended for weak or wayward growers.

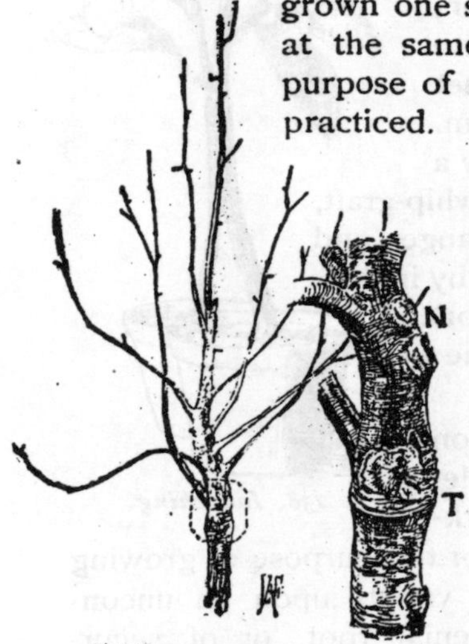

140. A double-worked tree.

Grafting Waxes.—There are great numbers of recipes for waxes or mastics for protecting grafts and covering wounds upon trees. In this country, the resin and beeswax waxes are most used, although some of the alcoholic waxes are popular in some regions. In Europe, many clay and pitch waxes are in common use. For most purposes, the wax No. 1, in the following list, will be found to be one of the best, especially for applying by the hand. The soft alcoholic waxes are apt to melt off exposed stubs in our hot summer suns; but they are useful for indoor work and for cool weather. In making the resin and beeswax waxes, the materials are first broken up finely and melted together. When thoroughly melted, the liquid is poured into a pail or tub of cold water. It soon becomes hard enough to handle, and it is then pulled and worked over until it becomes tough or "gets a grain," at which stage it becomes the color of very light-colored manilla paper. When wax is applied by hand, the hands must be well greased. Hard cake tallow is the best material for this purpose. In top-grafting large trees, it is well to carry a supply of tallow when waxing, by smearing the backs of the hands before entering the tree.

Common Resin and Beeswax Waxes.

1. Resin, 4 parts by weight; beeswax, 2 parts; tallow, 1 part.

2. Resin, 4 lbs.; beeswax, 1 lb.; tallow, 1 lb.

3. Resin, 6 lbs.; beeswax, 2 lbs.; linseed oil, 1 pt.

4. Resin, 6 lbs.; beeswax, 1 lb.; linseed oil, 1 pt. Apply hot with a brush, one-eighth of an inch thick over all the joints.

5. Resin, 4 lbs.; beeswax, 1 lb.; and from half to a pint of raw linseed oil; melt all together gradually, and turn into water and pull. The linseed oil should be entirely free from cotton-seed oil. A hard wax, for use in warm weather.

6. Resin, 6 parts; beeswax, 1 part; tallow, 1 part. To be used warm, in the house.

7. Resin, 4 or 5 parts; beeswax, 1½ to 2 parts; linseed oil, 1 to 1½ parts. For outdoor work.

Alcoholic Waxes.

8. Lefort's Liquid Grafting Wax, or Alcoholic Plastic.—Best white resin, 1 lb.; beef tallow, 1 oz.; remove from the fire and add 8 ounces of alcohol. Keep in closed bottles or cans.

9. Alcoholic Plastic with Beeswax.—Melt 6 parts white resin with 1 part beeswax; remove from stove and partially cool by stirring, then add gradually—with continued stirring—enough alcohol to make the mixture, when cool, of the consistency of porridge. In the temperature of the grafting-room it will remain sufficiently plastic to permit applying to the cut surfaces with the finger.

10. Alcoholic Plastic with Turpentine.—Best white resin, 1 lb.; beef tallow, 1 oz.; turpentine, 1 teaspoonful; add enough alcohol (13 to 15 fluid ounces of 95 per cent. alcohol) to make the wax of the consistency of honey. Or, less alcohol may be added if the wax is to be used with the fingers.

French and Pitch Waxes.

11. Common French.—Pitch, ½ lb.; beeswax, ½ lb.; cowdung, 1 lb. Boil together, melt, and apply with a brush.

12. Common French Bandage Wax.—Equal parts of beeswax, turpentine and resin. While warm spread on strips of coarse cotton or strong paper.

13. Grafting Clay.—⅓ cowdung, free from straw, and ⅔ clay, or clayey loam, with a little hair, like that used in plaster, to prevent its cracking. Beat and temper it for two or three days until it is thoroughly incorporated. When used it should be of such a consistency as to be easily put on and shaped with the hands.

14. Resin, 2 lbs. 12 ozs.; Burgundy pitch, 1 lb. 11 ozs.

At the same time, melt 9 ounces of tallow; pour the latter into the former while both are hot, and stir the mixture thoroughly. Then add 18 ounces of red ochre, dropping it in gradually and stirring the mixture at the same time.

15. Black pitch, 28 parts; Burgundy pitch, 28 parts; beeswax, 16 parts; grease, 14 parts; yellow ochre, 14 parts.

16. Black pitch, 28 lbs.; Burgundy pitch, 28 lbs.; yellow wax, 16 lbs.; suet or tallow, 14 lbs.; sifted ashes, 14 lbs. When used, warm sufficiently to make it liquid, without being so hot as to injure the texture of the branches.

17. Melt together 1¼ lbs. of clear resin and ¾lb. of white pitch. At the same time melt ¼lb. of tallow. Pour the melted tallow into the first mixture, and stir vigorously. Then before the stuff cools, add, slowly stirring meantime, ½lb. of Venetian red. This may be used warm or cold.

Waxed String and Bandage.

18. Waxed String for Root-grafting.—Into a kettle of melted wax place balls of No. 18 knitting cotton. Turn the balls frequently, and in five minutes they will be thoroughly saturated, when they are dried and put away for future use. This material is strong enough, and at the same time breaks so easily as not to injure the hands. Any of the resin and beeswax waxes may be used. When the string is used, it should be warm enough to stick without tying.

19. Waxed Cloth.—Old calico or thin muslin is rolled on a stick and placed in melted wax. When saturated it is allowed to cool by being unrolled on a bench. It is then cut in strips to suit. Or the wax may be spread upon the cloth with a brush.

Waxes for Wounds.

20. Any of the more adhesive grafting waxes are excellent for dressing wounds, although most of them cleave off after the first year. Stiff and ochreous paints are also good.

21. Coal-tar.—Apply a coating of coal-tar to the wound,

which has first been pared and smoothed. If the wound contains a hole, plug it with seasoned wood.

22. Hoskins' Wax.—Boil pine tar slowly for three or four hours; add ½lb. of beeswax to a quart of the tar. Have ready some dry and finely sifted clay, and when the mixture of tar and wax is partially cold, stir into the above named quantity about 12 ounces of the clay; continue the stirring until the mixture is so stiff and so nearly cool that the clay will not settle. This is soft enough in mild weather to be easily applied with a knife or spatula.

23. Schæfell's Healing Paint.—Boil linseed oil (free from cotton-seed oil) one hour, with an ounce of litharge to each pint of oil; then stir in sifted wood ashes until the paint is of the proper consistency. Pare the bark until smooth. Paint the wound over in dry weather, and if the wound is very large, cover with a gunny-sack.

24. Tar for Bleeding in Vines.—Add to tar about three or four times its weight of powdered slate or some similar substance. Apply with an old knife or flat stick.

25. Hot Iron for Bleeding in Vines.—Apply a hot iron to the bare surface until it is charred, and then rub into the charred surface a paste made of newly-burnt lime and grease.

26. Collodion for Bleeding in Vines.—It may be applied with a feather or small brush. In some extreme cases, two or three coats will be needed, in which case allow the collodion to form a film before applying another coat. Pharmaceutical collodion is better than photographic.

4. NURSERY MANAGEMENT.

The greater part of the field nurseries of the United States are engaged in raising grafted or budded plants. It is germane to the present chapter, therefore, to add some general notes upon the management of nurseries and nursery lands. A large part of the management of these establishments, however, is pure business, and is governed by the general laws of trade, and lies outside the field of the present discussion.

Nursery Lands.—The best land for general nursery purposes is one which is heavy rather than light, containing a good percentage of clay, and lying as nearly level as possible. Before trees are put upon it, the land should be deeply and thoroughly worked for at least one season, and if it is of such character as to hold surface water for two or three days at a time, the area should be thoroughly tile-drained. Nursery trees constitute a crop which occupies the land for a number of years, and unless this land is in good heart when the trees are planted, there will be little opportunity to raise a good product. With fruit trees, the age of the tree determines its salableness; hence it is imperative that the growth within the given time be rapid and strong. With ornamentals, however, the value is determined by the size of the specimen, with little reference to its age. It therefore follows that lands which are not sufficiently strong to allow of the profitable growing of fruit trees may still be useful for growing ornamentals. In considering the question of the fertility of nursery lands, it is first necessary to determine what are the proportions of the chief elements of plant food which the trees remove from the soil. Roberts (Bulletin 103, Cornell Experiment Station) gives the following figures upon this point:*

"Amounts and values of fertilizing constituents removed by an acre of nursery trees in three years:

	Apples.		Pears.		Peaches.		Plums.	
	Lbs.	Value.	Lbs.	Value.	Lbs.	Value.	Lbs.	Value.
Nitrogen	29.07	$4 36	24.83	$3 73	22.42	$3 36	19.75	$2 96
Phosphoric acid	10.13	71	7.83	54	5.42	38	4.42	31
Potash	19.73	89	13.33	60	11.75	53	11.50	52
		$5 96		$4 87		$4 27		$3 79

"The above results show conclusively that but a small amount of plant food is removed from the soil by the growth of nursery stock. They also show that more phosphoric acid is removed by the apples and pears than by

*See, also, 10th Rep. N. Y. State Exp. Sta., pp. 162-174.

the peaches and plums; but any ordinary soil, cultivated as nursery lands are, should easily furnish in three years ten times the plant food used by the trees. In order to compare the drafts made by nursery stock and some of the common crops raised in mixed husbandry, the following statement will be useful: The amount of green corn necessary to remove an equal amount of fertilizing ingredients per acre, taking the average of the value of the nitrogen, phosphoric acid and potash ($4.72) removed by an acre of the trees (3 years' growth), would be 4,779 pounds.

"Ensilage corn raised in drills usually yields from 12 to 20 tons per acre, and yet does not make drafts on the land which preclude duplicating the yield the following season; hence some other cause than soil exhaustion must be found if the failure to grow a second crop of nursery trees without intermediate crops is explained."

All experience proves that a crop of nursery trees does not exhaust the land of its fertility. In fact, it is generally considered that land from which trees have just been removed is in the very best condition for a crop of beans, wheat or potatoes. Yet, despite this fact, it is also generally considered that land can seldom raise two good crops of nursery trees in succession. Land which has been "treed" must be "rested" in grass or some other crop. This disposition of land to refuse to grow two consecutive crops of good trees is not an invariable rule, however. The writer has known nursery land to produce good plum trees for twenty consecutive years. One frequently sees lands producing apple and cherry stocks for two or three crops in succession. Plums seem to be particularly amenable to this consecutive cropping, and they are benefited by applications of stable manure. Some other species, as, for example, the pear, do not take so kindly to treatment with manure. Because of this common experience with indifferent trees grown upon treed land, nurserymen with a large business prefer to rent land for the growing of trees. In New York state, the common period of rental is five years, at a rate of about eight dollars per acre per year, for the ordinary type of farm lands.

The reason for this condition of treed lands is that the soil is injured in its physical texture by the methods of cultivation and treatment. The best nursery lands are those which contain a basis of clay, and these are the ones which soonest suffer under unwise treatment. The land is kept under high culture, and it is therefore deeply pulverized. There is practically no herbage on the soil to protect it during the winter. When the crop is removed, even the roots are taken out of the soil. For four or five years, the land receives practically no herbage which can rot and pass into humus. And then, the trees are dug in the fall, often when the soil is in unfit condition, and this fall digging amounts to a fall plowing. The soil, deeply broken and robbed of its humus, runs together and cements itself before the following summer; and it then requires three or four years of "rest" in clover or other herbage crop to bring it back into its rightful condition. This resting period allows nature to replace the fiber in the soil, and to make it once more so open and warm and kindly that plants can find a congenial root-hold in it. It would seem, therefore, that some of this mechanical injury to nursery lands might be prevented by the growing of some cover crop between the rows late in the season, to be plowed under the following spring. It is well known that the plowing-in of very coarse manure between the trees in fall or spring, for two or three years, will sometimes so greatly improve the land that a second good crop of trees can be grown upon the land with ease. This is particularly true for plum trees, as already noted, but the results do not seem to be so well marked for pears and some other trees. It is probable that one reason for the very general refusal of pear trees to follow pear trees is the fact that they demand heavy clay, and this is just the land which is most injured by nursery practices. Some lands are naturally so loose and open in structure that two or three crops of trees can be grown in succession, but these lands contain little crude clay, and therefore do not suffer quickly from the burning out of the humus.

Although the chemical analyses of nursery trees show comparatively small amounts of the more important plant foods, it may still often occur that nursery lands need fertilizing. Nitrogen is needed in comparatively large amounts. This is the element which chiefly conduces to strong growth. It is also the one which is most rapidly augmented by the addition of humus and the improvement of the physical condition of the soil, as recommended above. When nursery stock is making a poor growth, the grower should first see that the tillage of the soil is made as thorough and perfect as possible, in order to supply additional plant food and to preserve the soil moisture. He may then add nitrogen in the form of nitrate of soda or sulphate of ammonia, sowing them at the rate of 200 to 400 lbs. to the acre. The application should be made in spring or early summer. He should then be sure that insect or fungous attacks are averted. If the land was originally in fit condition for trees, and adapted to them, these suggestions should afford relief.

Grades of Trees.—Common opinion demands that a tree, to be first-class, must be perfectly straight and comely. This arbitrary standard is but the expression of the general demand for large and handsome trees. But there are some varieties of fruit trees which cannot be made to grow in a comely fashion, and hence there is always a tendency to discontinue growing them, notwithstanding the fact that they may possess great intrinsic merit. All this is to be deplored. The requirements of a first-class tree should be that the specimen is vigorous, free from disease or blemishes, and that it possess the characteristics of the variety. This allows a crooked tree to be first-class if it is a Greening or Red Canada apple, because it is the nature of these varieties to grow crooked. A crooked or wayward grower is not necessarily a weak one. It is advisable to top-work weak-growing varieties upon strong-growing and straight-growing ones (see page 134).

A first-class tree is well grown; that is, the various operations to which it has been subjected by the nurseryman have been properly performed. It must be mature; that is, not stripped of its leaves before the foliage has thoroughly ripened. It must be of the proper age for planting. It must have a clean, smooth bark. It must have a stocky, strong trunk, good roots, and be free of borers and other insect injuries. The union—at the bud or graft—must be completely healed over. Stocky and rather short trees, with well-branched heads, are always preferable to very tall ones. Very slender trees, if above one or two years old, should be avoided. Nurserymen express the size of a tree by its diameter about three inches above the bud. The measuring is usually done by a caliper. The diameter of a first-class tree varies with the method of growing and trimming it. In the New York nurseries, a first-class two-year-old apple tree (budded) should caliper five-eighths to three-fourths of an inch. Plums run about the same. Pears will generally run a sixteenth of an inch less, and sour cherries about a sixteenth more. Sweet cherries will run three-fourths inch and above. Nurserymen use various instruments for gauging the diameter of stock. The old-fashioned caliper is most commonly employed. An excellent modification of this device is the self-registering caliper, seen in Fig. 141. Heikes' tree-gauge, made of sheet steel, in shown in Fig. 142.

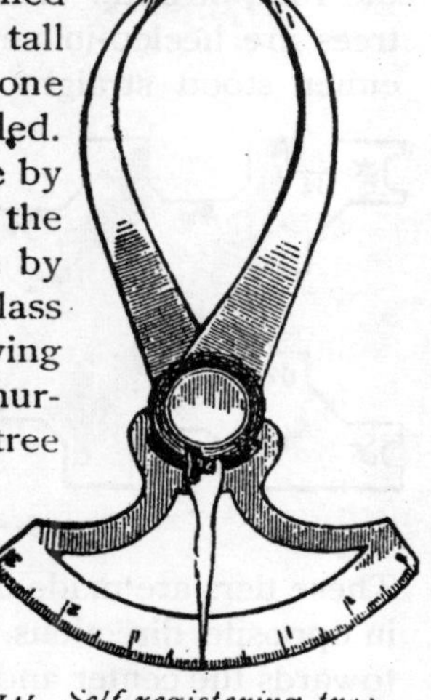

141 Self-registering tree-caliper.

The Storing of Trees.—Of late years, the nursery business has been greatly benefited by the free use of cellars for the storing of stock. In these cellars the stock is safe from winter injury, and it can be moved to customers before the nursery land is fit to dig in the spring. These cellars make the nurseryman somewhat independent of conditions of weather and trade, and they ensure to the planter quick delivery of stock which shows no winter injury. A common style of nursery cellar is shown in Fig. 143. It is a wooden structure, commonly a third or quarter below the surface of the ground, with hollow walls and a tarred and gravelled roof. It should be provided with ample facilities for ventilation, either by means of windows along the sides or flues in the roof, or both. It has a dirt floor. In this building, the trees are heeled-in very thickly in the fall. The trees are either stood straight up, or they may be piled in tiers.

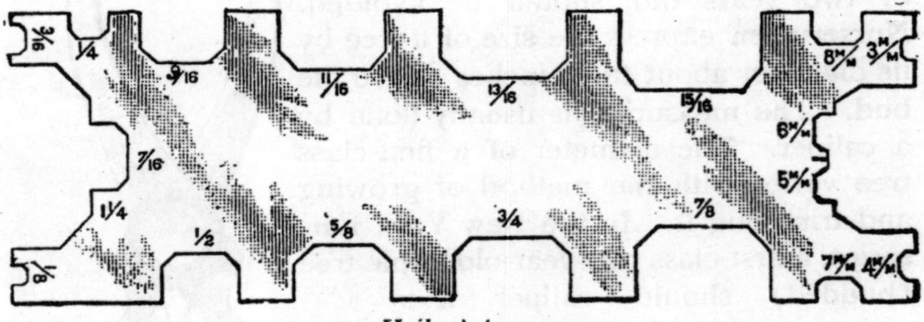

142. Heikes' tree gauge.

These tiers are made up of overlapping horizontal layers laid in opposite directions. The roots of the first layer are laid towards the center and damp sand thrown over them. Upon these are laid the roots of the second layer, with the tops in the opposite direction. Dirt is again thrown on, when another layer like the first is added. The tops are, therefore, always outward. These tops should lie a little higher than the roots, and in order to raise them, and also to bind the pile, scantlings or boards are laid crosswise of the layers, at the outward end, at intervals. Moss may be used in place of sand, although the latter is more easily obtained and kept, and is generally used. In piling or cording trees in this fashion, it is important that a sufficient passage or alley be left between each pile to admit of free circulation of air. A passage through which a man can just pass is sufficient.

A cellar a hundred feet long, twenty feet wide, and ten feet high in the clear, will winter about 25,000 three-year-old apple trees, if the trees are corded, as already described.

These storage cellars soon engender mold or fungus if they are allowed to become too warm or too close. Cellars with floors as high as the surface of the ground keep "sweeter" than those which are sunken. The remedy for this fungus, which often does great damage to stock, is to keep the house well aired, and then to kill it out by fumigating. A common practice is to burn shavings or sawdust in the cellar, and then open the doors and windows and air the place. If the smudge is dense, the fungus is said to be easily destroyed. Evaporating sulphur—not burning it—upon an oil stove is also effective. Place the sulphur in a

143. Storage cellar.

pan and set this pan in another of about the same size, in the bottom of which is a layer of sand a half inch thick. Place both of them upon the stove, and allow the sulphur to melt and evaporate, filling the house with the fumes. The layer of sand will prevent the sulphur from catching fire, unless it is allowed to run over. Burning sulphur quickly kills all plants which are in active growth. Its action upon dormant nursery stock is unknown to the writer. A low temperature and an abundance of fresh air, however, are the best safeguards against fungus. They are also essential to the preservation of the bright, vivid color of the stock. Trees which are wintered in close and warm cellars look dull in the spring. The temperature should be kept as near freezing as possible. When the stock is not being handled, a slight frost does no damage.

In heeling-in trees in the open for the winter, care should be exercised to select a well-drained and protected place. The roots are placed in furrows and covered, and the tops are laid down almost horizontal. Another row is lapped over the first, much as shingles are lapped over each other. Loose straw or litter about the place should be removed or tramped down, else mice may rest in it and girdle the trees. An excellent device to keep mice out of a heeling-in yard is to place a foot board on edge all about the place, leaning the top out a little. Hold the boards in place by stakes, close up the cracks, and tramp the earth against the bottom of the boards, and the mice are completely fenced out. If it is necessary to cover the tops of peach and other tender trees, evergreen boughs will be found to be the most satisfactory protection.

Trimming Trees in the Nursery.—One of the chief efforts of the nurseryman is to make his trees stocky. Many factors conspire to produce this result. Any treatment which makes trees grow vigorously may be expected to contribute to their stockiness, if the grower does not circumvent it by some subsequent operation. The trees should be given plenty of room. The rows in the nursery should stand 3½ feet apart, for ordinary fruit trees, and the plants should stand 10 inches or a foot apart in the row. During this first year, the leaves should not be rubbed off the bodies of the trees, else the trees will grow too much at the top and become too slender. If, however, strong forking or side branches appear low down — as often happens in sour cherries—they should be removed. Budded stock should reach a height of 4 feet or more the first year. The following spring, the stock is headed-in uniformly, reducing it to the height of 3 or 4 feet, according to kind and the uses for which the stock is grown. In New York nurseries, the average apple stock is headed back to a height of about 3 feet 3 inches to 3 feet 5 inches. Sweet cherries are headed 2 to 3 inches taller. Sour cherries are generally not headed-in, because they make a less tall growth; but if they go much above 3 feet they are headed back. Soon after the trees are headed back this second spring, they are "sprouted." This operation consists in hoeing the dirt away from the base of the tree and cutting off all sprouts which start from the root or the crown. After heading-in, the tree "feathers out" from top to bottom. It is a common practice to rub off these new shoots which appear upon the body, allowing only those shoots to remain which spring from near the top of the trunk, and which are presumed to form the top of the future tree. This rubbing off of the side shoots early in the second season is generally to be condemned. It tends to make the tree grow top-heavy, whilst the body remains spindling and weak. A better plan is to allow the shoots to remain until July or early August, when they are cut off close to the trunk. The wounds will then heal over, or nearly so, by fall, and the tree will have grown strong and stocky.

Dwarfing.—The dwarfing of trees depends upon two factors,—working upon a slow-growing stock, and subsequent heading-in. In particular cases, dwarfing is also accomplished by growing the trees in pots or boxes. The nurseryman supplies the first factor,—the tree united to the dwarf root. But this factor alone rarely insures a permanently dwarf tree. The vigorous top will soon impart some of its habit to the stock; and if the tree is planted so deep that the union is a few inches below ground, roots may start from the cion, and the tree will become half-dwarf, or even full standard. The capability of keeping the tree dwarf lies mostly with the grower, although, unfortunately, the grower usually ascribes it wholly to the nurseryman. An excellent illustration of all this is afforded by the cherry. If cherry trees are to be dwarfed, they are to be worked upon the Mahaleb cherry; and yet the greater part of the sweet cherries, and some of the sour ones, are budded upon Mahaleb roots in eastern nurseries, but our cherry trees are not dwarfs thereby. If, however, the grower were to head-in his Mahaleb-worked cherries each year, in the same way as he is advised to treat his dwarf pears, he would be able to have dwarf trees. In like manner, the plum upon the Myrobalan plum, the peach upon the plum, the apple upon the Doucin or even upon the Paradise, soon cease to be dwarfs if allowed to grow to their utmost. The pear upon the quince affords the most complete dwarf fruit tree which we have, but even this soon ceases to be a true dwarf if heading-in is neglected.

There are many varieties of plants which are dwarf by nature, and they therefore do not require to be worked upon slow-growing stocks. The Paradise apple is itself such a natural dwarf, and was originally a seedling. (For an account of dwarf apples, see Lodeman, Bulletin 116, Cornell Experiment Station.) Dwarf spruces, pines, viburnums, beans, dahlias, and scores of other plants are well known. Such dwarfs are generally propagated by means of cuttings, although some of them, as the garden vegetables and annual flowers, reproduce themselves from seeds. The particular methods of dealing with these varieties are detailed under the respective species in the next chapter.

Root-grafted vs. Budded Trees.—There has been a most controversial discussion of the relative merits of root-grafted and budded fruit trees these many years. For the most part, this discussion has been unprofitable, for there has been litttle earnest effort to arrive at any just or exact method of comparison. The disputants have too often dealt in generalized statements, and it must be said that prejudice, and the desire to advocate the particular stock which one is growing, are not unknown to these discussions. Some experiments have been tried for the purpose of determining the relative merits of the two methods of propagation, but none of the experimenters seem to have really analyzed the subject or to have arrived at any truthful conclusions. We must approach the subject in an analytical spirit if we are to hope for useful results.

Before proceeding to a discussion of the comparative effects of budding and root-grafting, it is essential that certain definitions be clearly fixed in the mind. The budding of fruit-stocks in the nursery is performed in the summer time upon stocks which were set in the spring, as fully explained on pages 94 to 105. These stocks are trimmed or "dressed" before they are set in the nursery. Root-grafting, as already explained (See Figs. 103, 104), is the setting of a cion upon a root. If the entire root is used, the operation is known as whole-root-grafting. In this case, the cion is set at the crown and the root is dressed in much the same way that the stock is dressed when it is to be used for budding. If only a portion of the root is used as stock (as in Fig. 103), the operation is known as piece-root-grafting. It is this particular operation which is ordinarily understood when people speak of root-grafting. It is apparent that the various pieces made of the root may not be comparable. The top piece includes the crown, at which point the cion is inserted. The lowest piece comprises the tip, or smallest, and therefore weakest, portion of the root. Ordinarily, about three pieces are made of a root in the root-grafting of apple stocks.

It is evident that there are two distinct problems concerned in the consideration of the comparative merits of budded and root-grafted trees. One has to do with the comparison of the budding with the grafting, and the other with the different methods of trimming or cutting the stocks. It is perfectly well known that, in general, budding and grafting are equally efficacious methods of propagation, other things being equal. In other words, the mere fact that one tree comes from a bud and another from a cion should make no necessary difference in the value of the tree. All the characteristic differences between budded and root-grafted trees are due to the methods of trimming the stocks, and not to the actual methods of propagation.

It is indisputable that there is great difference in the root system between the ordinary budded tree and the ordinary root-grafted tree. The roots of the root-grafted tree, as it leaves the nursery, are comparatively shallow and horizontal, and are generally prongy and strongly developed on one side or another of the tree. It is well known, of course, that different varieties of apples develop a different root system in the nursery row, but the same variety ordinarily has a very different root development when propagated by budding and by common root-grafting. The writer has seen this difference so uniformly for so many years, and upon such an extent and variety of stock, both east and west, that he has no hesitation in positively affirming that, as generally grown, the root system of budded trees is unlike that of root-grafted trees.

This difference in root development proceeds from the method of cutting the stock. In other words, if the pieces of roots were budded they would undoubtedly develop the same system of roots that they do when grafted. The philosophy of it will become apparent upon a moment's reflection. The short piece of root has fewer side rootlets than the whole or long root. It is these side rootlets which develop into the main branches of the root system. The root system of the piece-root must, therefore, be shallower at first start than that of the whole root, because the axis is shorter. Moreover, these side rootlets do not develop simultaneously upon all sides of the main axis. They are scattered along the axis. A section or piece of the root may contain rootlets only on one or two sides of the axis, and as these rootlets grow the system becomes one-sided. There is still another reason for the prongy and one-sided character of the root-system of piece-roots. The piece of

144. New roots on the end of a piece-root.

root is essentially a cutting. Every gardener knows that roots seldom start symmetrically from all sides of the end of a cutting. Fig. 144 (from a photograph) shows young roots springing off from the end of a cutting. All three of them start from nearly a common point. It is a one-sided or unsymmetrical system. Fig. 145 shows two root-grafts, drawn from life, as they had grown at the expiration of two months after they were planted in the nursery. They show the same peculiarities of root development as the cutting does in Fig. 144.

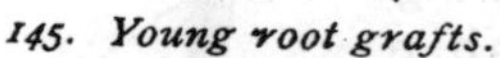
145. Young root grafts.

The reader now desires to know why the same one-sided method of root growth does not take place at the end of the root in the budded tree, for these stocks are dressed or trimmed—that is, the tips of the roots are cut off—before they are set in the nursery row. The whole question turns upon how much the roots of the stocks are cut back. If only the very tip is cut off, and there is a strong root development above it, this tip will simply heal over and develop no side roots, or else what side roots do develop will be very weak. This is practically what takes place in the common treatment of budding stock. If, however, the root were very severely cut back, the same development would no doubt start from the tip of the budded stock as from that of the root-grafted stock. Fig. 146, from life, shows how this may occur. The stock on the left is budded, that on the right grafted. Both were severely headed-in (cut off at T), and both have developed prongy roots. The budded stock was much longer than the other, however, and, therefore, its root system is stronger.

The whole question, therefore, is one of comparative length and strength of roots (or stocks). A whole-rooted tree should be stronger and have a more symmetrical root system, at a given age, than a piece-rooted tree. Yet there have been frauds committed in the name of whole-rooted trees. As a matter of fact, there can be no perfectly whole-rooted trees unless the bud or cion is set upon a seedling stock which stands in its original position, for some of the main root axis is broken off in the process of digging. Yet, if stock is well dug, this shortening-in of the tip of the root is so slight as to be practically of no account.

146. Budding and grafting on piece-roots.

If the pieces of roots are very short in the making of root-grafts, the graft has too little power to enable it to make a strong growth the first year. It is a very common practice to cut off the entire top of the root-grafted tree at the end of the first year, in order to get a strong and straight body the following year. This practice is perfectly justifiable only that the grower counts the age of his tree from the date of the cut-back, and not from the date of the grafting. Root-grafted trees are very likely to make such a short growth the first season that if the terminal bud should be winter-killed, the tree will branch too low, or if a

147. Root-graft, headed back

leader starts from a lateral bud the body will be crooked. A good nurseryman always wants his first season's growth to be high enough to form the entire body of the tree. If this body is obliged to grow on from its terminal bud the second season, the annual ring can be plainly seen on the body—an indisputable mark of age, which the customer will be quick to discern. Fig. 147, from life, shows a common method of dealing with root-grafted trees. The union is at A, and the top of the original cion at B. At the end of the first season (or the following spring), the tree was cut

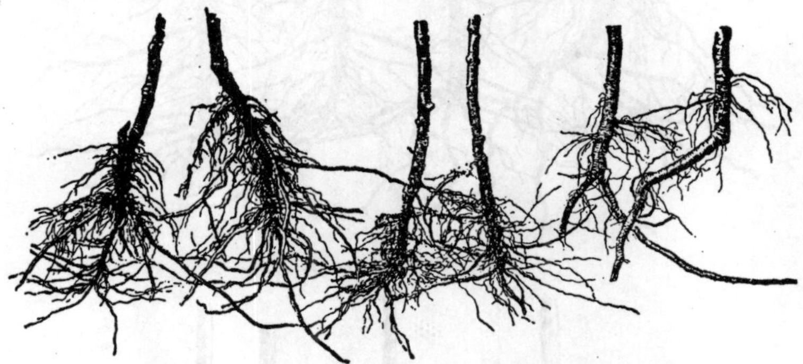

148. Ben Davis trees, budded and root-grafted.

back to C. The nurseryman will count the age of his tree from the point C.

At the same actual age, and grown in the same place, the budded tree is nearly always larger than the root-grafted tree, as ordinarily grown. The longer and better the piece of root upon which the graft is made, however, the less the difference will be. The illustrations, all from actual and typical trees, show some of these differences. Fig. 148 shows six Ben Davis apple trees grown in a New York nursery. The two trees upon the left are budded. The other four are root-grafted. The two middle trees had been transplanted, but the two upon the right stood where the grafts were planted. It will be seen how completely the transplanting has broken up the tendency to tap-roots and prongs, and has developed a more symmetrical root system. The root system of the budded trees is deeper and more symmetrical because the stocks or roots were longer. Figs. 149 and 150 each show, beginning at the left, Fallawater, Golden Russet, Hubbardston and Gravenstein apple trees.

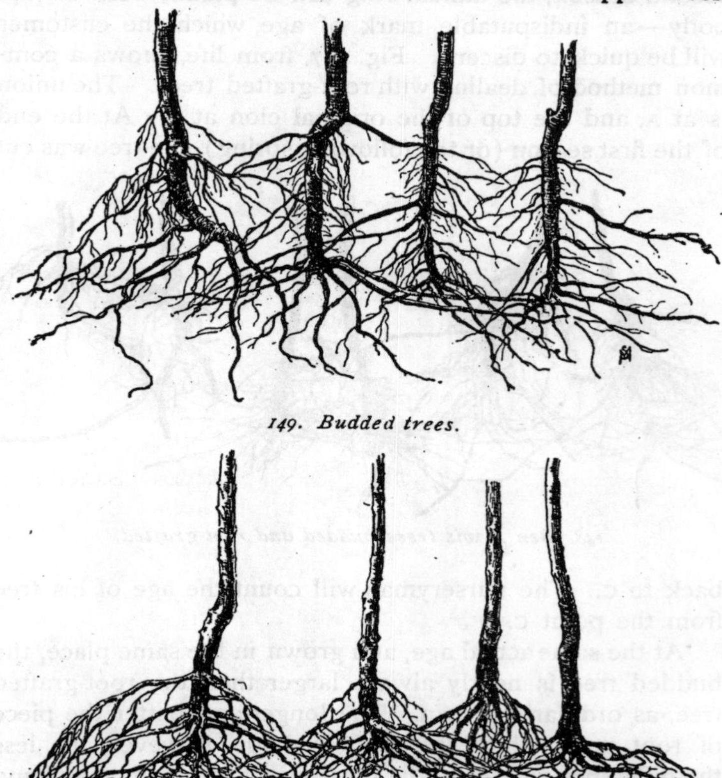

149. Budded trees.

150. Root-grafted trees.

Those in Fig. 149 are first-class three-year budded trees from an eastern nursery. Those in Fig. 150 are first-class three-year root-grafted trees from a western nursery. The disparity in sizes of short-piece-root trees and budded trees of like actual age, is well seen in Figs. 151 and 152. They are Mann apples. In Fig. 151, the piece-root-grafts, upon the left, are two years from the graft; the buds, upon the right, are of like age. In Fig. 152, the piece-root-grafts, upon the left, are three years old, and the buds, upon the right, are two years. The different root systems of the two are apparent in each case.

All these comparisons are not made for the purpose of showing that root-grafts are inferior to buds, but simply that they are different from them. Yet, the author is convinced that very many of the root-grafted trees are made with such short and weak pieces of roots that the trees are distinctly inferior. The practice of root-grafting fruit trees has almost disappeared from the east. Eastern buyers generally desire strong, heavy trees, with deep and full root systems; and there is an opinion—though not resting upon definite experiments — that the deep-rooted budded trees enter deeper into the ground and make longer-lived trees than the root-grafted samples.

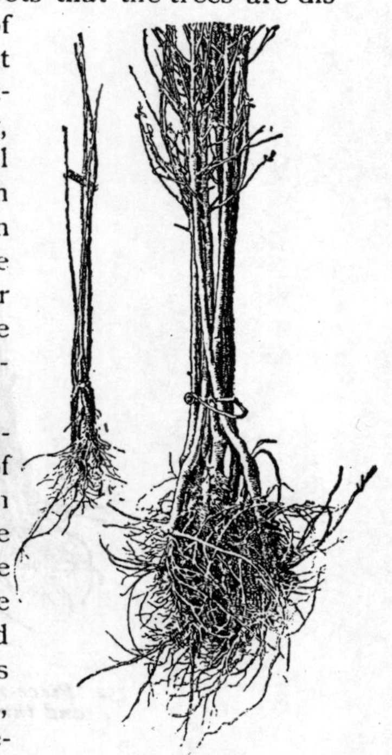

151. Piece-root-grafts and buds, two years old.

The entire question of the ultimate merits of the two classes of trees rests, therefore, more upon the way in which the stocks are trimmed and handled when the propagating is done, than upon the mere fact of their being budded or root-grafted. Root-grafting has distinct merits in the northwest, where own-rooted trees are desired (see Fig. 104), and it cheapens propagation; but as propagating is ordinarily done in our nurseries, the author is distinctly of the opinion that, as a rule, the budded apple tree is a stronger and better tree, as it leaves the nursery, than the root-grafted tree is when of the same age and when grown under the same conditions. He is equally convinced, on the other hand, however, that it is possible to grow as good trees by root-grafting as by budding.

NOTE.—The student, who may desire to pursue the subject of graftage further, should procure Charles Baltet's "L'Art de Greffer." There is an English edition.

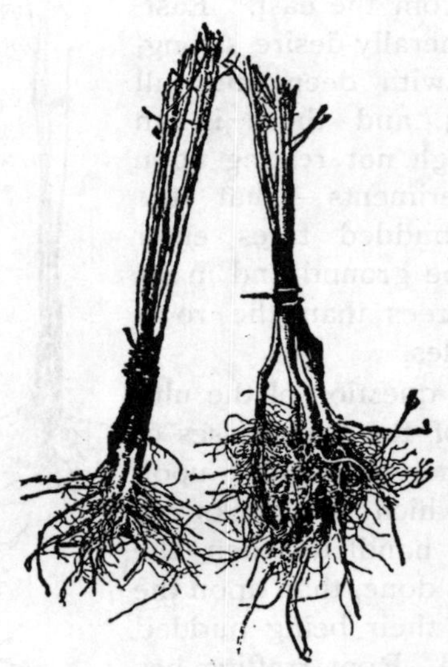

152. Piece-root-grafts and buds, two and three years respectively.

CHAPTER VI.

THE NURSERY LIST.

Aaron's Beard. See Hypericum.

Abelia. *Caprifoliaceæ.*

In spring by layers under a frame, and in summer by cuttings.

Abies (Fir). *Coniferæ.*

Propagated by seeds, which are usually kept dry over winter and sown in spring in frames or in protected borders. Cones should be fully matured before being gathered. If they hold the seeds tightly they should be placed in a dry place, sometimes even in an oven, until the scales spread. In some species, as the Balsam fir, the cones drop and fall to pieces as soon as ripe, and these cones must be gathered just before they begin to fall. The seeds may be separated by rubbing them in the hands, when they are thoroughly dry, then winnowing them out through a sieve. In order to obtain stocky plants, the seedlings should be transplanted the following spring. The named varieties and the species which do not produce sufficient seed are winter-worked upon seedling stocks which are potted in the fall. Cuttings of growing tips set in sand in a close, well-shaded house or frame are often successful. (See Figs. 47 and 67, and page 64.) Stocks the size of a lead pencil are commonly used. One-year-old seedlings are usually preferred, but in some cases the requisite size is not reached until the second or third year. Any of the common operations of grafting may be employed, but the veneer-graft is best. The conifers are not difficult to graft. The European Silver fir (*Abies pectinata*) or the Balsam fir may be used as a stock, but the common Norway spruce is now the most popular stock for species of both Abies and Picea (see Picea).

Abobra. *Cucurbitaceæ.*

Propagated by seeds, or rarely by soft cuttings.

Abroma. *Sterculiaceæ.*

By seeds sown in March. By cuttings made in spring from half-ripened wood, and placed under a bell-glass.

Abronia (Sand Verbena). *Nyctaginaceæ.*

Propagated by seeds sown in autumn or spring, after the outer skin has been peeled off. Sow in pots of sandy soil, and keep in a frame until the following spring; then place in their flowering quarters. By young cuttings, set in spring, in sandy soil.

Abrus. *Leguminosæ.*

Propagated by seeds raised in heat or by cuttings under a hand-glass, in sand.

Abutilon. *Malvaceæ.*

Sow seeds in pans, with same soil and temperature as for cuttings. By cuttings from young wood, at almost any season; the best time, however, is spring or fall. Insert in pots, in a compost of equal parts peat, leaf mold, loam and sand, and place in a temperature of 65° to 75°.

Acacia. *Leguminosæ.*

Propagated by seeds sown as soon as ripe, in sandy peat, about one-fourth inch deep, or a little more for large seeds. Soak in hot water 24 hours if seeds are not fresh. Keep temperature about 55° or 60°, and pot off when large enough to handle. By cuttings of the half-ripened wood, put in with a heel, in equal parts peat and sand, covered with pure sand. Insert the cuttings as soon as made; water, and leave them in the shade till dry. Place under a bell-glass, shade and water to prevent flagging. Pot off when rooted, and keep in a close pit or house until the plants are thoroughly established. *A. pubescens* and some others will strike from root-cuttings. See, also, Robinia.

Acalypha. *Euphorbiaceæ.*

Propagated by cuttings in sandy soil under a glass, in stove heat, during late winter or in spring. Native species by seeds.

Acanthephippium. *Orchidaceæ.*

Propagated by dividing the pseudo-bulbs as soon as growth commences. (See under Orchids.)

Acanthopanax. Like Aralia.

Acanthophœnix. *Palmaceæ.*

Propagated by seeds, sown in a moist bottom heat, in a well decomposed compost of one part loam, one of peat, one of leaf mold, and one of sand.

Acanthus (Bear's Breech). *Acanthaceæ.*

Propagated by seeds sown in gentle heat, or by division of the root in autumn or early spring. Also by root-cuttings. Water carefully.

Acer (Maple). *Sapindaceæ.*

Stocks are grown from stratified seeds, which should be sown an inch or two deep. Some very early-ripening species, as *A. dasycarpum* and *A. rubrum* (the silver or soft maple and the red maple), come readily if seeds are simply sown as soon as ripe. They will not keep well until the next spring. Varieties are often layered, but better plants are obtained by grafting. The Japanese sorts are winter-worked on imported *A. polymorphum* stocks, either by whip- or veneer-grafting. Varieties of native species are worked upon common native stocks. Maples can also be budded in summer, and they grow (generally with some difficulty) from cuttings of soft and ripe wood.

Aceras. *Orchidaceæ.*

Propagated by carefully made divisions of the tubers. (See under Orchids.)

Achillea. Including Ptarmica (Yarrow, Milfoil). *Compositæ.*

Propagated by seeds, root divisions and cuttings, during spring.

Achimenes, including Scheeria. *Gesneraceæ.*

Propagated by seeds, carefully sown in well-drained pans, which are filled nearly to the rim, leveled, and well watered with a fine rose. Sow seed and cover lightly with sand, and place in a shady position. Keep moist, and apply water very lightly. Place a sheet of glass over the seed-pan. After large enough to be pricked off, treat like rooted cuttings. By scales of the corms, rubbed off and sown like seeds, barely covered with sand, and placed in bottom heat. By leaves, set into pots of similar soil as for cuttings, placing all the petiole below the surface, and placed in bottom heat. (Fig. 81.) By cuttings from any portion of the stem ; insert in a soil of equal parts of peat and sand, in well-drained pots, in bottom heat.

Achras. See Sapota.

Achyranthes. See Iresine.

Aconitum (Aconite, Monk's Hood, Wolf's Bane). *Ranunculaceæ.*

Seeds should be sown as soon as ripe in a coldframe or border ; also by division. Roots should not be left about, for they are very poisonous.

Acorns. See Quercus.

Acorus. *Aroideæ.*

Propagated during spring by divisions.

Acrophyllum. *Saxifragaceæ.*

Increased by cuttings of the half-ripened shoots, which strike freely in a soil of sand and peat ; cover with a hand-glass, and place in a cool house. The roots should be kept moist.

Acrostichum. See Ferns.

Actæa (Baneberry). *Ranunculaceæ.*

Propagated by seed and by division of roots during spring.

Actinidia. *Ternstrœmiaceæ.*

Propagated by seeds, layers or cuttings. The cuttings should be put in sandy soil, in autumn, under a handlight.

Actinotus. *Umbelliferæ.*

Increased by seeds sown on a hotbed in spring, and in May the seedlings may be transplanted to the open border in a warm situation, where they will flower and seed freely. Divisions of the roots grow readily.

Ada. *Orchidaceæ.*

Propagated by divisions as soon as the plant commences growth. (See under Orchids.)

Adamia. *Saxifragaceæ.*

Increased by seeds ; by cuttings, which will root readily in a compost of loam, peat and sand, under a hand-glass.

Adam's Needle. See Yucca.

Adenocarpus. *Leguminosæ.*

Seeds may be sown in March, the hardy species outdoors, and the others in a cold house. Young cuttings root freely in sand, if covered with a hand-glass.

Adenophora. *Campanulaceæ.*

Propagated by seeds, sown as soon as ripe, in pots placed in coldframes. Also by suckers.

Adenostoma. *Rosaceæ.*

Propagated by cuttings of the young shoots, placed in sand, under glass, in spring or autumn.

Adina. *Rubiaceæ.*

Propagated by cuttings placed in rich, loamy soil, under a hand glass, in heat.

Adlumia (Allegheny Vine, Smoke Vine, Mountain Fringe). *Fumariaceæ.*

Propagated by seeds. The plant is a biennial, blooming the second season only.

Adonis. *Ranunculaceæ.*

Propagated by seeds. The perennials may be divided at the root.

Ægle (Bengal Quince, *Citrus trifoliata*). *Rutaceæ.*

Propagated by ripe cuttings, which will root in sand under a hand-glass, in heat, if not deprived of any of their leaves. Also by seeds.

Ærides. *Orchidaceæ.*

The only method of propagating this genus is by removing the upper portion and planting it separately. It should always be severed low enough to include a few roots, otherwise a large proportion of leaves will be lost. A somewhat dense shade, a moist atmosphere and careful watering are essential until the young plant is established. The old stool will soon send out lateral growths, which, in time, may be separated and treated similarly. Vanda, Saccolabium, Angræcum, Renanthera, are increased in the same way. (See under Orchids.)

Æschynanthus. *Gesneraceæ.*

Propagated by seeds, which are very unsatisfactory. By cuttings, which root freely in a well-drained pot, filled with a light compost, and having a surface of pure white sand, about one inch deep, during spring. The best are obtained from half-ripened wood, cut into two or three-inch lengths, and all leaves, with the exception of one or two at the top, removed. Cover the cuttings with a bell-glass, and place in moderate bottom heat. When rooted, transfer singly to small pots, place under hand-glasses until thoroughly established, then gradually harden off.

Æschynomene. *Leguminosæ.*

Propagated by seeds, those of the herbaceous species requiring a good heat to start them into growth. By cuttings, placed in sand under a bell-glass, in a brisk heat.

Æsculus (Horse Chestnut, Buckeye). *Sapindaceæ.*

Propagated by stratified seeds sown in single rows in spring, and by layers made in the spring or fall; or by grafting or budding on the common horse chestnut or native buckeye, usually under glass.

Aganisia. *Orchidaceæ.*

Propagated by dividing the pseudo-bulbs just before starting into new growth. (See under Orchids.)

Agapanthus (African Lily). *Liliaceæ.*

Propagated by offsets, or by divisions of the old plants in early spring.

Agaricus. See Mushroom.

Agathæa (species of Felicia). *Compositæ.*

Seeds and layers. Young cuttings root freely, in a gentle heat, at all times.

Agathosma. *Rutaceæ.*

Increased by cuttings, which, when young, root freely in a pot of sand, under a bell-glass, in a cool house. They require to be shaded somewhat in the summer.

Agati. *Leguminosæ.*

Increased by cuttings, which will root in a pot of sand with a hand-glass over them, placed in heat.

Agave. *Amaryllidaceæ.*

Increased by seeds, to secure the production of which the flowers generally need to be pollinated. Usually by suckers, which spring naturally from the old plant.

Ageratum, Cælestina. *Compositæ.*

Sow the seeds in January, in heat, in sandy soil. When large enough, prick them off into thumb pots, and keep in heat till they grow freely; then place them in a cooler house. Cuttings are commonly used for propagation.

Agrostis (Bent Grass). *Gramineæ.*

Increased easily by seeds, sown in spring in the open border.

Ailanthus (Tree of Heaven). *Simarubaceæ.*

Propagated by suckers; and by pieces of the roots planted in a pot with their points above the ground, and placed in a hotbed. Seeds are used when large quantiites are desired.

Ajuga (Bugle). *Labiatæ.*

Perennials, propagated by seeds sown in the open border, during spring or autumn; by divisions. Annuals, by seeds.

Akebia. *Berberidaceæ.*

Seeds. Layers of young or ripe wood. Dormant (or firmwood) cuttings, under glass in summer.

Albuca. *Liliaceæ.*

Propagated by seeds and by offsets (bulbels) from the old bulb.

Alchemilla (Lady's Mantle). *Rosaceæ.*

Propagated by seeds or by divisions of the roots.

Alder. See Alnus.

Aleurites. *Euphorbiaceæ.*

Propagated by ripe cuttings in sand, under a handglass. Do not remove leaves.

Alexandrian Laurel. See Ruscus.

Alfalfa. See Medicago.

Algaroba Bean. See Carob.

Alhagi (Manna Tree). *Leguminosæ.*

Increased by seeds sown in a hotbed; and by cuttings rooted in sand, with a bell-glass over them, in heat.

Alisma (Water Plantain). *Alismaceæ.*

Increased by seeds, which should be sown in a pot

immersed in water and filled with loam, peat and sand; also by divisions, which root well in a moist, loamy soil.

Allamanda. *Apocynaceæ.*

Layers. Shoot cuttings will root well at any time of the year in a bottom heat of from °70 to 80°. The usual time is, however, in spring, when the old plants are pruned back. Choose the tops of the shoots, retaining two or three joints to each cutting. Place in a compost of sand and peat or leaf-mould in equal proportions, singly, in pots, and plunge the pots in the propagating bed.

Allium, including Porrum, Schœnoprasum. *Liliaceæ.*

Increased by seeds sown thinly in light soil in early spring. By bulbels, planting them in autumn or spring 1 to 4 inches deep. See Onion, Leek, Chives, Garlic.

Allosorus. See Ferns.

Almonds (*Prunus Amygdalus, P. Japonica*). *Rosaceæ.*

The almond is worked the same as the peach and apricot. Seedling almond stocks are best, but the peach is often used. Apricot stocks are sometimes employed, but they are not to be recommended.

Double-flowering almond will grow from root-cuttings if on own roots. Heel-in plants in fall, and buds will begin to form in three to six weeks; then make cuttings. Results are poor when cuttings are taken directly upon first lifting the plants. See Prunus.

Alnus (Alder). *Cupuliferæ.*

Propagated usually by seeds, which are gathered in the fall and well dried. Then they are sprinkled lightly on the ground and covered very thinly. Towards the end of the year the seedlings are planted in rows 1½ feet apart, and 6 inches from each other, where they may remain for two years, after which they can be placed where they are intended to stand. Planting is best done in October or April. They are also increased, but rarely, by suckers, by cuttings and by grafting.

Alocasia. *Aroideæ.*

Increased by seeds and divisions, as for caladium.

Aloë. *Liliaceæ.*

Commonly propagated by suckers, which spring from the base of the plant. Seeds are sometimes employed.

Alonsoa. *Scrophulariaceæ.*

Propagated by seeds, sown in spring; also by cuttings in sandy soil, in gentle heat. The herbaceous species may be treated as outdoor summer annuals, and should be raised in little heat, and planted out in May.

Aloysia (species of Lippia, Sweet-scented Verbena, Lemon Verbena). *Verbenaceæ.*

Increased easily in spring by young wood. The cuttings will root in about three weeks, in sandy soil with gentle heat. Also by cuttings of ripened wood in autumn.

Alsophila. See Ferns.

Alstrœmeria. *Amaryllidaceæ.*

Increased by seeds. By a careful division of the fleshy roots, during fall or spring.

Alternanthera (*Telanthera Bettzichiana*). *Amarantaceæ.*

Commonly raised from cuttings of growing wood. For spring and summer bedding, the plants are started in late winter. The stock plants, from which cuttings are taken, are procured from cuttings made late in summer. Seeds are little used.

Althæa (Marsh-Mallow, Hollyhock). *Malvaceæ.*

Increased by seeds, and by divisions. The biennial species must be raised from seeds every year. See Hollyhock.

Alum Root. See Heuchera.

Alyssum (Madwort). *Cruciferæ.*

Increased by seeds (particularly sweet alyssum and other annuals) sown in the open border or in pans of sandy soil. By divisions and layers. By cuttings made from young shoots two to three inches in length, placed in sandy loam, early in the season, in a shady place.

Amarantus. *Amarantaceæ.*

Propagated by seeds sown in hotbeds in spring, and thinned out when about one-half inch high. Late in spring they may be transplanted outdoors in their permanent situation, or into pots. Sometimes sown in the open.

Amaryllis. *Amaryllidaceæ.*

Propagated by seeds and offsets. Seedlings will bloom in from one to two years.

Amelanchier (Shad-bush, Juneberry, Service berry). *Rosaceæ.*

Seeds. Layers and cuttings in autumn. By grafting, in early spring, on the mountain ash, hawthorn or the quince, or the weaker on the stronger-growing species. See Juneberry.

Amellus. *Compositæ.*

Increased by divisions; or by cuttings under glass in spring.

Amherstia. *Leguminosæ.*

Propagated by seeds; also by cuttings of the half-ripened wood inserted in sand under a glass, in bottom heat of about 80°.

Amianthemum. See Zygadenus.

Amorpha (Lead Plant, Bastard Indigo). *Leguminosæ.*

Increased by seeds, usually. Layers or cuttings, taken off at the joint, strike readily if placed in a sheltered situation early in autumn. They should remain undisturbed till the following autumn.

Amorphophallus. *Aroideæ.*

Propagated by offsets, or cormels, and by seeds, which, however, are usually sparingly produced in cultivation.

Ampelopsis. *Vitaceæ.*

Increased by seeds, especially the one known as *A. Veitchii*, or Boston ivy (properly *A. tricuspidata*). Layers or cuttings made in spring from the young soft wood, root freely in gentle heat. By cuttings having a good eye, if taken in September and pricked under hand-lights in sandy soil on the open border, or in pots. Hard-wood cuttings or rooted runners are commonly employed in this country for *A. quinquefolia* (Virginia Creeper).

Amphicome. *Bignoniaceæ.*

Increased by seeds, sown in early spring, in pots of sandy soil placed in a greenhouse. By young shoots inserted in sandy soil in gentle heat in spring.

Amsonia. *Apocynaceæ.*

Propagated by seeds; by divisions of the roots in spring; or by cuttings during the summer months.

Amygdalus. See Prunus.

Anacardium (Cashew). *Anacardiaceæ.*

Ripened cuttings, with their leaves left on, root freely in sand under a hand-glass, in heat.

Anagallis (Pimpernel). *Primulaceæ.*

The annuals, by seeds sown in a warm place in spring; the perennials, by cuttings from young shoots, or by division, at any time, either under a hand-glass or in a closed frame. Keep in the shade, and when thoroughly established harden off gradually.

Ananas. See Pine Apple.

Anantherix. *Asclepiadaceæ.*

Increased by seeds, which ripen in abundance, or by division of the root.

Anastatica (Resurrection Plant). *Cruciferæ.*

Increased by seeds sown in the spring in heat, and the plants afterwards potted off and plunged again in heat to hasten their growth.

Anchusa. *Borraginaceæ.*

Propagated by seeds, which should be sown in early spring in pots of sandy soil; they will germinate in three or four weeks. Also by divisions, and rarely by cuttings.

Andersonia. *Epacridaceæ.*

Propagated by cuttings from tips of young shoots. These should be made in autumn, winter or spring, and planted in sand in a gentle heat, with a bell-glass over them.

Andromeda. *Ericaceæ.*

Propagated by seeds, sown thinly as soon as ripe, in pots or pans, in sandy peat soil. Living sphagnum is an excellent material upon which to sow andromeda seeds. Place in a cool frame or greenhouse, giving plenty of air. The young plants should be planted out in spring, if large enough, or pricked into boxes if small. By layers, which, if carefully pegged down during September, will take twelve months to make sufficient roots to allow of their being separated; layerage is a common method.

Anemia. See Ferns.

Anemone (Wind Flower). *Ranunculaceæ.*

Propagated by seeds, root divisions or root cuttings in autumn or early spring; the seeds are better sown as soon as ripe in pans, in a coldframe.

Angelica. *Umbelliferæ.*

Increased by seeds, which should be sown in September or March, in ordinary soil.

Angelonia. *Scrophulariaceæ.*

Propagated by seeds, which should be planted in spring in hotbeds, and transplanted in the open in May. By cuttings of the young shoots in spring. These root readily under a hand-glass or in a propagating-bed, if given plenty of air daily.

Angræcum. See Ærides.

Anguloa. *Orchidaceæ.*

Propagated by dividing the pseudo-bulbs, just before they commence to grow. (See under Orchids.)

Anisanthus. See Antholyza.

Anise. *Umbelliferæ.*

Increased by seeds sown in ordinary soil, on a warm, sunny border in spring.

Anœctochilus. *Orchidaceæ.*

Propagated by cutting off the growing top just below the last new root, dividing the remainder of the stem into lengths of two or three joints. (See under Orchids.)

Anomatheca (referred by some to Lapeyrousia). *Iridaceæ.*

Increased sometimes by seeds sown very thinly in seed pans as soon as ripe. Also, multiply very rapidly by cutting up the masses once a year. Offsets.

Anona (Custard Apple). *Anonaceæ.*

Increased by seeds, which, in the north, should be sown in pots and plunged into a hotbed. By ripened cuttings, which will root in sand under a hand-glass, in a moist heat.

Ansellia. *Orchidaceæ.*

Increased by divisions of the tubers just after flowering. (See under Orchids.)

Antennaria. *Compositæ.*

Propagated by seeds sown in spring in a coldframe, and by divisions of the roots in spring.

Anthemis (Chamomile). *Compositæ.*

Propagated by seeds and divisions.

Anthericum, Phalangium. *Liliaceæ.*

Increased by seeds sown as early as possible after they are ripe, in a coldframe; by division of the roots.

Antholyza, including Anisanthus. *Iridaceæ.*

Increased by seeds, which should be sown as soon as ripe, in light soil, in a cool house. Here they will germinate the following spring, and will be fit to plant out in the summer of the same year. Also by offsets.

Anthurium. *Aroideæ.*

Propagated by seeds sown as soon as ripe in shallow, well-drained pans or pots filled with a compost of peat, loam, moss, broken crocks or charcoal, and clean sand. Cover lightly and place in a close, moist propagating case, where a temperature of 75° to 85° is maintained; or the pots may be covered with bell-glasses. Keep the soil in a uniformly moist condition. Also increased by divisions, which should be made in January.

Anthyllis (Kidney Vetch). *Leguminosæ.*

Herbaceous perennials, increased by seeds or cuttings. The cuttings of most species will root in a pot of sandy soil, with a bell-glass over them, in a cool house or frame. Seed of the annuals should be sown in a warm, dry place in the open ground.

Antirrhinum (Snapdragon). *Scrophulariaceæ.*

Increased by seeds sown in early spring or midsummer; by cuttings, which should be taken in September, when they will readily root in a coldframe, or under a hand-glass.

Aphelandra. *Acanthaceæ.*

Propagated by cuttings from half-ripened wood taken off with a heel. Cut the base of each clean across; insert an inch apart in pots of sandy soil, and plunge in a brisk bottom heat.

Apios (Ground-Nut). *Leguminosæ.*

Propagated by the tubers, or divisions of them; also easily by seeds.

Aplectrum (Putty-Root). *Orchidaceæ.*

Increased by the bulb-like subterranean tubers; also by seeds. A difficult plant to grow.

Apocynum (Dog's Bane). *Apocynaceæ.*

Propagated by seeds, suckers and divisions. The best time to divide is just as the plants are starting into growth in spring.

Aponogeton. *Naiadaceæ.*

Increased rapidly by seeds and offsets. The seeds should be sown as soon as ripe, in pots plunged in water and covered with glass.

Apple (*Pyrus Malus*). *Rosaceæ.*

Standard apple stocks are grown from seeds, and dwarf stocks from mound layers. Apple seeds are either imported from France or are obtained from pomace. The French seeds give what are technically known as *crab stocks*, the word *crab* being used in the sense of a wild or inferior apple. The yearling stocks themselves are imported from France in great numbers. It has been supposed that French crab stocks are hardier and more vigorous than ours, but this opinion is much less common than formerly, and the foreign stocks are not so popular now as the domestic stocks. As a rule, nurserymen who grow trees do not raise apple stocks. Stock growing is largely a separate business, and in this country it is an important industry in Kansas, Nebraska, Iowa, and other plains states.

The chief source of apple seeds at the present time is the pomace from cider mills. The "cheese" of pomace is broken up, and if the material is dry enough it may be run through a large sieve to remove the coarser parts. The seeds are then removed by washing. Various devices are in use for washing them out. They all proceed upon the fact that the pomace will rise in water and the seeds sink. Some use a tub or common tank, which is tilted a little to allow the water to flow over the side. Others employ boxes some 7 or 8 feet long, 4 feet wide and a foot deep, the lower end of which is only 11 inches deep to allow the escape of the water. This

Apple, continued.

box is set upon benches, and a good stream of water is carried into it at the upper end. A bushel or two of pomace is emptied in at a time, and it is broken and stirred with a fork or shovel. When the seeds are liberated, they fall to the bottom and the refuse runs over the lower end. Another box is provided with several cleats, at intervals of about a foot, and the ends are left open. The box is set at an angle, and the seeds are caught behind the cleats. Seeds must not stand long in the pomace pile, or they will be seriously injured. Nurserymen like to secure the pomace as soon as it is taken from the press.

As soon as the seeds are collected, they should be spread upon tables or boards, and should be frequently turned until perfectly dry. They may then be stored in boxes in slightly damp sand or sawdust, or in powdered charcoal, and kept in a cool and dry place until spring. Or if they are to be sown immediately, they need not be dried, but simply mixed with enough dry sand to absorb the water so as to make them easy to handle. Seeds should not be allowed to become hard and dry through long exposure to the air, or they will germinate unevenly. Apple seeds procured at the seed stores are often worthless because of this neglect. Very dry seeds can sometimes be grown, however, by subjecting them to repeated soakings, and then sprouting in a gentle hotbed or mild forcing-house. Change the water on the seeds every day, and at the end of a week or ten days mix with sand and place in a thin layer in the hotbed. Stir frequently to prevent molding. When the seeds begin to sprout, sow them in the open ground. This operation, which is sometimes called *pipping*, may be performed in a small way near the kitchen stove. Seeds are sometimes "pipped" between moist blankets. (See also page 17.)

When sowing is done in the fall, the seeds may be sown in the pomace. This entails extra labor in sowing, but it saves the labor of washing. This practice gives good results if the pomace is finely broken, and it is now common among nurserymen.

In loose and well-drained soils, sowing is undoubtedly best performed in the fall, just as early as the seeds are ready. But upon land which holds much water, and which heaves with frost or contains much clay, spring sowing is preferable. In spring, the seeds should be sown just as soon as the ground can be worked.

Apple, continued.

If the stocks are to be cultivated with a horse, the rows should be 3 or 3½ feet apart. Some growers sow in narrow drills and some in broad ones. The broad drills are usually 6 to 10 inches wide. The earth is removed to the depth of 2 or 3 inches, if it is loose and in good condition, the seed is scattered thinly on the surface and the earth hoed back over them. If the ground is likely to bake, the seeds should not be sown so deep; and it is always well, in such cases, to apply some very light and clean mulch. The plants should be well cultivated during the season, and they should attain a height of 6 to 12 inches or more the first year. If the plants come thickly, they must be thinned out.

In the fall of the first year the seedlings should be large enough to be dug and sold to general nurserymen. Sometimes the poorest plants are allowed to stand another year, but they are usually so scattering that they do not pay for the use of the land, and they should be transplanted the same as the larger stock, or the weakest ones may be thrown away. The stocks are dug with a plow or tree-digger and heeled-in closely, so that the leaves "sweat" and fall off. The plants are then stored in sand, moss or sawdust in a cellar. Before they are **planted.** the tops are cut off near the crown, usually with a hatchet on a block. The stocks are then graded into budding and grafting sizes. The general nurserymen buy these stocks in fall or early winter. Those which are root-grafted are worked during late winter, but those intended for budding, or which must be grown another season before they attain sufficient size for working, are heeled-in, sometimes being "dressed" (see Chapter V); in the spring they are set in nursery rows, about a foot apart in the row (page 146). The nurseryman reckons the age of his tree from the top or graft, rather than from the time the seed was sown.

Seedling raising is usually conducted by men who make it a business, and who supply the general nurserymen of the country. It is largely practiced at the west, where the deep and strong soils produce a rapid growth. The yearling trees are graded by the western growers into about four lots: "Extras," or those at least ¼ inch in diameter at the crown, and having 12 inches of both top and root; these are used mostly as budding stocks the next season. "Commons," those between $\frac{3}{16}$ and ¼ inch at the crown, and having 8 inches of root; these are used for immediate

Apple, concluded.

root-grafting. "Second-class," those from $\frac{2}{16}$ to $\frac{3}{16}$ inch at the crown, and "third-class," or all those under $\frac{2}{16}$. The last two classes must be grown in the field for one or two seasons before they can be worked to advantage.

Dwarf stocks are mostly obtained from mound-layering. The common stock for dwarfing is the Paradise apple, a dwarf variety of the common apple species (*Pyrus Malus*). This variety rarely attains a height of more than 4 feet. A larger or freer stock is the Doucin, also a variety of *Pyrus Malus*, which will produce an engrafted tree intermediate in size between that given by the Paradise and free or common stocks. This is little used in this country. To obtain stools for mound-layering, the tree, when well established, is cut off within 4 or 6 inches of the ground in spring, and during the summer several shoots or sprouts will arise. The next year the stool is covered by a mound, and by autumn the layers are ready to take off. Sometimes, when stocks are rare, mound-layering is performed during the first summer, before the young shoots have hardened, but good stocks are not obtained by this method. Common green layering is sometimes practiced the first year, but it is not in favor. The dwarf stocks, in common with all apple stocks, may be sparingly propagated by root-cuttings and by hard-wood cuttings.

Apple stocks are either grafted or budded. Root-grafting is the most common, especially at the west, where long cions are used in order to secure own-rooted trees. (See Figs. 103, 104.) Budding is gaining in favor eastward and southward; it is performed during August and early September in the northern states, or it may be begun on strong stocks in July by using buds which have been kept on ice. Stocks should be strong enough to be budded the same year they are transplanted, but the operation is sometimes deferred until the second summer. Stocks which cannot be worked until the second year are unprofitable, especially on valuable land. For root-grafting, strong one-year-old roots are best, but two-year-olds are often used. (See pages 148 to 156.)

In common practice, the root is cut into two or three pieces of 2 to 3 inches each, but stronger trees are obtained, at least the first year or two, by using the whole root and grafting upon the crown. The lowest piece is usually small and weak, and is generally discarded.

The apple is easily top-grafted and top-budded. (See Chapter V. For grades of trees, see page 142.)

Apricot (*Prunus Armeniaca, P. dasycarpa, P. Mume*). *Rosaceæ.*

The apricot thrives upon a variety of stocks. Apricot stocks are used in apricot-growing regions, especially for deep and rich, well-drained soils. The pits grow readily if given the same treatment as that detailed for the peach (which see). The stocks are also handled in the same manner as peach stocks. Apricots upon apricot roots are not largely grown outside of California, in this country. Apricot stocks can be grown from root cuttings the same as cherries and other stone fruits.

The apricot does well upon the peach, especially on light soils. In the warmer parts of the country peach is much used.

Plum stocks are commonly used at the north, especially if the trees are to be planted in moist or heavy soils. The common plum is generally used, but some of the native plum stocks are now coming into favor, especially in trying climates. The Russian apricots, which are a hardy race of *Prunus Armeniaca*, are grown in colder climates than the common varieties, and they therefore demand hardy stocks. Any of the native plums make good stocks, but the Marianna is now coming into especial prominence. The myrobolan plum can be used for all apricots, but it is not popular, particularly in severe climates. (See Bulletin 71, Cornell Experiment Station.)

The almond, both hard and soft-shelled, is sometimes used for the apricot, but the union is likely to be imperfect, and it is not recommended. Almond-rooted trees are thought to be best adapted to light soils.

Varieties of apricots are usually budded, in the same way as the peach, although they may be side-grafted at the crown in the nursery row.

Aquilegia (Columbine, Honeysuckle erroneously). *Ranunculaceæ.*

Increased by seeds. They must be sown very thinly, soon after being ripe, in a sandy soil or in pans in a coldframe. Division of the root is the only way to perpetuate any particular variety with certainty.

Arabis (Wall Cress, Rock Cress). *Cruciferæ.*

Increased by seeds sown in the border or in pans, in spring; by divisions of the root, and by cuttings placed in a shady border during summer.

Arachis (Pea-Nut, Goober, Ground-Nut). *Leguminosæ.*

Increased by seeds, which, for greenhouses or cold climates, should be sown in heat; and, when the plants have grown to a sufficient size, they should be potted off singly. The peanut, as a field crop, is grown from seeds planted where the crop is to stand.

Aralia, Dimorphanthus. *Araliaceæ.*

Propagated by seeds and by root cuttings; also by stem cuttings, in heat. See Ginseng.

Araucaria. *Coniferæ.*

Increased by seeds sown in pans or boxes, with but gentle heat. By cuttings from the leading shoots, placed firmly in a pot of sand; they first require a cool place, but afterwards may be subjected to a slight warmth. When rooted, pot off into fibrous loam, mixed with leaf soil and sand.

Arbor-vitæ. See Thuya.

Arbutus (Strawberry Tree). *Ericaceæ.*

Increased by seeds, which should be sown in sand during early spring, and by grafting, budding or inarching upon *A. Unedo.*

Arbutus, Trailing. See Epigæa.

Ardisia. *Myrsinaceæ.*

Propagated by seeds and cuttings.

Areca (Cabbage Palm). *Palmaceæ.*

Increased by seeds, which should be sown in a compost of loam, peat and leaf soil, in equal parts, with a liberal addition of sand, and placed in a moist and gentle heat.

Arenaria (Sandwort). *Caryophyllaceæ.*

Increased by seeds, division or cuttings; the last placed under a hand-glass will root freely. Seeds should be sown in spring in a coldframe. The best time to divide the plant is early spring, or during July and August.

Argemone. *Papaveraceæ.*

Increased by seeds, which may be sown outdoors in spring, those of the rarer species in a hotbed.

Argyreia (Silver-weed). *Convolvulaceæ.*

Propagated by cuttings, which will do well in sand, with a hand-glass over them, in a little bottom heat.

Arisæma. Consult Arum.

Arisarum. *Aroideæ.*

Propagated in spring by seeds or divisions of the root.

Aristea. *Iridaceæ.*

Increased by seeds and divisions.

Aristolochia (Brithwort). *Aristolochiaceæ.*

Propagated by seeds and layers, which are not very satisfactory. Cuttings of tender sorts root freely in sand, with bottom heat. The seeds must be fresh.

Armeria (Thrift, Sea Pink). *Plumbaginaceæ.*

Increased by seeds sown in spring, in pots of sandy soil, and placed in a coldframe; by division, separate pieces being planted as cuttings under hand-glasses.

Arnebia. *Borraginaceæ.*

Seeds. Cuttings of the strong shoots should be inserted in pots of sandy soil, and placed in gentle heat.

Arnica. *Compositæ.*

Propagated by seeds sown in a coldframe in spring, and by divisions, which should be made in spring.

Arrow-root. See Calathea.

Artabotrys. *Anonaceæ.*

Propagated by seeds; and in the north by cuttings of ripened wood, placed in early spring in sand under a frame, with bottom heat. Similar treatment to Anona.

Artemisia (Mugwort, Southernwood, Wormwood). *Compositæ.*

The annuals by seeds; the herbaceous ones by dividing at the root; the shrubby kinds by cuttings.

Arthrostemma. *Melastomaceæ.*

Propagated by cuttings of small, firm side shoots, which will root, in April or August, under a hand-glass in sandy soil.

Artichoke (*Cynara Scolymus*). *Compositæ.*

Grown from seeds. Although the plant is perennial,

a new stock should be started about every other year. It is increased also by suckers or divisions of the stools.

Artichoke, Jerusalem (*Helianthus tuberosus*). *Compositæ.*

Commonly increased by means of the tubers, which may be planted whole or cut into eyes, after the manner of potatoes. Seeds are very rarely used.

Artocarpus (Bread Fruit). *Urticaceæ.*

Propagation is difficult, as the plant is grown in northern countries. Suckers may be utilized when procurable. The young and slender lateral growths are used for cuttings.

Arum. *Aroideæ.*

Propagated by seeds, but usually by division of the roots, the best time being just as they begin their new growth, securing as many roots as possible to each division. Any rootless pieces should be placed in heat shortly after removal; this hastens the formation of roots and excites top growth. Arisæmas are treated in the same way.

Arundinaria. *Gramineæ.*

Increased by division of the root.

Arundo (Reed). *Gramineæ.*

Propagated by seeds or divisions, spring being the best time for either method. In early autumn, the canes can be cut into lengths of 18 to 24 inches for cuttings, and partly buried in sand in a gentle bottom heat, laying them horizontally.

Asarum. *Aristolochiaceæ.*

Propagated easily by divisions in spring.

Asclepias (Milkweed, Silkweed). *Asclepiadaceæ.*

Increased by seeds sown in pots in spring, pricked out singly when large enough, and treated like cuttings. By cuttings, which should be secured in spring, struck in gentle heat, under a bell-glass, and as soon as they are well-rooted potted into small pots. Seeds of *A. tuberosa* must be sown or stratified as soon as gathered.

Ascyrum. *Hypericaceæ.*

Increased by seeds and by careful divisions of the roots in spring.

Ash. See Fraxinus.

Asimina. *Anonaceæ.*

Propagated by seeds. The seedlings may be raised in pots, and sheltered carefully. By layers made in autumn.

Asparagus. *Liliaceæ.*

The common kitchen-garden asparagus is best propagated by means of seeds. These are sown in spring as soon as the ground can be worked, usually in rows a foot or two apart. Thin the young plants to 2 or 3 inches apart in the row and give good culture, and the plants can be set in the field the following spring, and they will give a fair crop after growing there two seasons. Small growers nearly always buy plants of nurserymen. Old asparagus crowns can be divided, but seeds give better plants.

The ornamental species of asparagus are propagated by seeds when they are obtainable; otherwise, by division. See Myrsiphyllum.

Asperula. *Rubiaceæ.*

Increased by seeds and by divisions of the roots during spring and early summer.

Asphodeline. *Liliaceæ.*

Propagated by division.

Asphodelus (Asphodel). *Liliaceæ.*

Propagated by seeds and by division of the root in early spring.

Aspidistra. *Liliaceæ.*

By division of the crowns, or by suckers.

Aspidium. See under Ferns.

Asplenium. See under Ferns.

Aster (Aster, Starwort, Michaelmas Daisy). *Compositæ.*

Propagated by seeds sown in spring, or by root divisions made in autumn; also by cuttings, which root freely in sandy soil under a hand-glass, with little heat. For China Aster, see Callistephus.

Astilbe. *Saxifragaceæ.*

Propagated by division in early spring, and by seeds if they are produced.

Astragalus (Milk Vetch). *Leguminosæ.*

Seeds should be sown in pots of sandy soil placed in a coldframe, as soon as ripe, or early in the spring, as they may lie a long time before germinating. The herbaceous perennials also increase by divisions, and the shrubby kinds slowly by means of cuttings placed in a coldframe.

Astrocaryum, Phœnicophorum. *Palmaceæ.*

Increased by seeds sown in spring in a hotbed; or by suckers, if obtainable.

Astroloma. *Epacridaceæ.*

Propagated by young cuttings placed in sandy soil, under a bell-glass, in a cool house.

Atalantia. *Rutaceæ.*

Propagated by ripened cuttings, which will root freely in sandy soil under a hand-glass, in heat.

Atamasco Lily. See Amaryllis.

Atragene (species of Clematis). *Ranunculaceæ.*

Seeds should be stratified, and sown in early spring, in gentle heat. By layering in autumn; the layers should not be separated for about a year, when they will be vigorous plants. By cuttings, which should be set in light soil and placed under a hand-glass.

Atriplex. See Orach.

Atropa (Belladonna). *Solanaceæ.*

Seeds.

Aubrietia. *Cruciferæ.*

Propagated by seeds, which should be sown in spring. In early autumn carefully transplant to a cool, shady border. Also by divisions. Where a stock of old plants exists, layer their long, slender branches any time after flowering, and cover with a mixture of sand and leaf soil; they will then root freely and establish themselves in time for spring blooming. Cuttings should be "drawn" or grown in a frame until they are soft, before they are removed.

Aucuba. *Cornaceæ.*

Readily increased by seeds, sown as soon as ripe; or by cuttings, inserted in spring or autumn in sandy soil, with or without a covering. The plant is tender at the north.

Auricula (*Primula Auricula*). *Primulaceæ.*

Propagated by seeds, sown as soon as ripe or in spring, in well-drained pots filled with sandy soil, well watered previous to sowing. Cover lightly with coarse sand, place a pane of glass over the pot, and place the latter in a hand-glass. By offsets, which should be removed when top-dressed, as they are more likely to root., Arrange about four offsets around the sides of well-drained 3-inch pots, filled with sandy soil, place under a bell-glass or in a close hand-light, and water very sparingly so as to prevent them damping off.

Australian Feather-palm. See Ptychosperma.

Averrhoa. *Geraniaceæ.*

Increased in spring by half-ripened cuttings, which will root in sand, under a hand-glass, with bottom heat.

Azalea. *Ericaceæ.*

Increased by seeds, sown as soon as ripe, or early the following spring, in a large, shallow frame containing from 2 to 3 inches of peat, over which more peat must be spread by means of a fine sieve; do not cover, but water thoroughly. Live sphagnum also makes an excellent soil. When the seedlings begin to appear they should have air, shade, and a daily sprinkling of water; transplant in autumn in boxes of peat and coarse sand, water, shade, and keep close until growth commences. Grafting is largely practiced to increase the stock of named varieties or choice seedlings, the stock most employed being *A. Pontica* for hardy sorts, and some strong-growing variety of *A. Indica*, like Phœnicia, for tender ones. Layering in spring, enclosing the part buried with moss, is also practiced; but the layer must be left two years before separating. Cuttings of *A. Indica* made of the hardened wood 2 or 3 inches long, taken with or without a heel, root readily in sand; about the end of summer is the best time. When placed outside they should be covered with a hand-light for about two months, and at the end of that time air should be given freely. See Rhododendron.

Babiana. *Iridaceæ.*

Propagated quickly by seeds sown in pans, placed in a gentle heat. These will grow at almost any time. The young plants will require to be carefully transplanted each season until they develop into blooming corms. By offsets grown in boxes or planted out in light, rich soil until large enough for flowering.

Backhousia. *Myrtaceæ.*

Increased by half-ripened cuttings, in sand, under a bell-glass, in a cool house, during spring.

Bactris. *Palmaceæ.*

Increased by suckers, which are very easily produced.

Bæa, Dorcoceras. *Gesneraceæ.*

Propagated easily by seeds.

Bæckea. *Myrtaceæ.*

Increased by cuttings of young wood, which will root freely if placed in a pot of sand, with a bell glass over them, in a cool house.

Bald-Cypress. See Taxodium.

Balm (*Melissa officinalis*). *Labiatæ.*

Seeds sown outdoors in spring. Division.

Balsam (*Impatiens Balsamina, I. Sultani*, etc.). *Geraniaceæ.*

Increased by seeds sown in early spring, in pans of rich, sandy soil, and placed in a gentle bottom heat of about 65°. Or the seeds may be sown directly in the garden when the weather becomes warm. Varieties increased by layers in late summer, under glass, or by veneer-grafting; also by cuttings. *I Sultani* is better raised from seeds than from cuttings. The stove species are multiplied by seeds, or cuttings in close frames.

Balsamodendron. *Burseraceæ.*

Increased by cuttings taken from the ripe young wood, in spring, and placed under a hand-glass, in bottom heat.

Balsam-tree. See Clusia.

Bambusa (Bamboo). *Gramineæ.*

Propagated by careful division of well-developed plants, in early spring, just as new growth is commencing; establish the divisions in pots. If young shoots are layered, leave only the end exposed.

Banana and **Plantain** (*Musa sapientum*, *M. paradisiaca* and others). *Scitamineæ.*

Edible bananas rarely produce seeds. The young plants are obtained from suckers, which spring from the main rootstock. These suckers are transplanted when 2 or 3 feet high. These plants themselves do not produce so good crops as the suckers which arise from them, and are not transplanted. Two or three suckers are sufficient for a plant at a time; what others arise should be transplanted or destroyed. The suckers should be set deep, as low as two feet for best results. In fifteen or eighteen months the plants will bloom, if they have had good care. The stem bears fruit but once, but new stems arise to take its place. See Musa.

Baneberry. See Actæa.

Banksia. *Proteaceæ.*

Seeds are very unsatisfactory. Propagated by well-ripened cuttings taken off at a joint, and placed in pots of sand without shortening any of the leaves, except on the part that is planted in the sand, where they should be taken off quite close. The less depth the better, so long as they stand firm. Place them under hand-glasses in a propagating house, but do not plunge them in heat.

Baptisia. *Leguminosæ.*

Increased by seeds, which should be sown in sand and leaf-mold in the open, or in pots placed in a coldframe. By divisions.

Barbadoes Gooseberry. See Pereskia.

Barbarea (Winter Cress, American Cress, Upland Cress). *Cruciferæ.*

Increased by seeds (chiefly), divisions, suckers and cuttings.

Barberry (*Berberis vulgaris*, etc.). *Berberidaceæ.*

Propagated by stratified seeds, or by suckers, layers and cuttings of mature wood. Layers are usually allowed to remain two years. Rare sorts are sometimes grafted on common stocks.

Barkeria (species of Epidendrum). *Orchidaceæ.*

Propagated by divisions made just before new growth commences. See under Orchids.

Barleria. *Acanthaceæ.*

Propagated by cuttings made of the young wood, and placed in a compost of loam and peat with a little rotten dung, under a bell-glass, in stove temperature with bottom heat.

Barrenwort. See Epimedium.

Bartonia aurea. See Mentzelia.

Basil (*Ocymum Basilicum* and *O. minimum*). *Labiatæ.*

Seeds, sown in a hotbed or outdoors.

Basswood. See Tilia.

Batatas. See Ipomæa and Sweet Potato.

Batemannia. *Orchidaceæ.*

Increased by divisions and offsets.

Bauhinia (Mountain Ebony). *Leguminosæ.*

Propagated by cuttings, which should be taken when the wood is neither very ripe nor very young The leaves must be dressed off, and the cuttings planted in sand under a glass in moist heat. Also by seeds.

Bayberry. See Myrica.

Bean. *Leguminosæ.*

Seeds ; sow only after the weather is thoroughly settled for outdoor culture. Lima beans should not be sown till a week or ten days after it is safe to sow the common kinds.

Bean Caper. See Zygophyllum.

Bean, Sacred or **Water.** See Nelumbo and Nymphæa.

Bear's Grass. See Yucca.

Beaucarnea. *Liliaceæ.*

Increased chiefly by seeds, which have been imported from their native country. By cuttings, when obtainable.

Beaufortia. *Myrtaceæ.*

Propagated by cuttings of half-ripened shoots ; place in a sandy soil under a glass, with very little heat.

Beech. See Fagus.

Beefwood. See Casuarina.

Beet (*Beta vulgaris*). *Chenopodiaceæ.*

Seeds, sown very early, before frosts cease for the early crop.

Befaria. *Ericaceæ.*

Propagated by cuttings of young wood, placed in sandy soil, in gentle heat.

Begonia. *Begoniaceæ.*

Increased by seeds, well ripened before they are gathered, and kept very dry until sown. For the successful raising of begonias, it is necessary to sow the seeds in pans or pots of well-drained, light, sandy soil, which should be well watered before the seeds are sown. The seeds should not be covered with soil, or they may fail to germinate. Place a pane of glass over the pans (Fig. 2), and set in a warm house or frame, where a temperature of about 65° can be maintained, and shade from the sun. As soon as the plants are large enough they should be pricked off into pans of light leaf-mold soil, in which they may remain until large enough to be placed singly in pots.

By divisions of the rhizomes. Also increased by cuttings, which strike freely in pots of sand and leaf-mold, and placed on a bottom heat of about 70°. Where large quantities are required, a bed of cocoanut fiber in a stove or propagating-frame may be used, and in this the cuttings may be planted, and remain until well rooted. Leaf cuttings of the Rex or foliage types are in common use. They succeed best when laid on sand or cocoanut fiber, and shaded from bright sunlight. Select old, well-matured leaves, and make an incision with a sharp knife across the principal nerves, on the under side. They should then be placed on the sand or fiber, and held down by means of a few pieces of crock. Under this treatment plantlets will form on the lower ends of the nerves of each section of the leaf, and these, when large enough, may be removed from the bed and potted. Fan-shaped pieces of leaves are often used. Leaf cuttings of begonia are described and figured in Chapter IV. (Figs. 78, 79, 80.) Species like *B. diversifolia*, etc., may be propagated by the tubers which form in the axils of the leaves. *B. phyllomaniaca* produces plantlets on the leaves and stems, and these may be removed and handled like small seedlings.

Tuberous Begonias (By E. G. Lodeman). — Tuberous

Begonia, continued.

species may be propagated by seeds, cuttings, and by divisions of the tubers. The seeds should be sown early in spring, and the seedlings pricked off and shifted as described for the evergreen or shrubby sections. Cuttings of the young, rapidly-growing shoots, if taken as soon as the plants are 4 to 6 inches high, will form good tubers by fall. Cuttings made while the plants are in flower rarely produce tubers of much value; *B. Boliviensis*, *B. Sedeni* and *B. Veitchii* are particularly apt to fail in this respect. The cuttings should be from 2 to 4 inches in length, the lower cut being just beneath a joint; remove one or two of the lowest leaves and insert singly near the edge of thumb-pots filled with a soil composed of about equal parts sand, leaf-mold and loam. Place in a cool, shaded position, applying water only to prevent flagging. Dividing the tubers is an unsatisfactory method of propagation, except in the case of *B. Socotrana*. The tubers should be cut before active growth begins, so that each part shall have an eye or crown. They are then treated as separate tubers. Begonias which have not been improved are most easily and rapidly propagated from seed; the named or improved varieties are best increased by cuttings.

Belamcanda, including Pardanthus (Blackberry Lily). *Iridaceæ.*

Seeds, division, and cuttings of young growth. The Blackberry Lily (*B. Chinensis*) propagates freely by division and by seeds.

Bellflower. See Campanula.

Bellis (Daisy). *Compositæ.*

Increased by seeds, which should be sown in early spring. By division after flowering, each crown making a separate plant. The soil must be pressed firmly about them.

Bellwort. See Uvularia.

Bengal Quince. See Ægle.

Bent Grass. See Agrostis.

Benthamia. *Cornaceæ.*

Propagated by seeds sown when ripe in a cool house, or by layering in autumn. By cuttings, and by grafting on the dogwood.

Benzoin. See Lindera.

Berberidopsis. *Berberidaceæ.*

Propagated by seeds in spring, by layering in autumn, or by young cuttings in spring.

Berberis. See Barberry.

Berchemia. *Rhamnaceæ.*

Propagated by layering the young shoots. By ripened cuttings, and slips of the roots planted under glass.

Bertolonia. *Melastomaceæ.*

Propagated by seeds and cuttings.

Bessera. *Liliaceæ.*

Propagated by offsets.

Betonica. See Stachys.

Betula (Birch). *Cupuliferæ.*

Increased by seeds, which must be sown as soon as gathered, or else stratified. By grafting or budding upon seedling stocks of the common kinds; the former should be done in spring or late winter, and the latter in summer when the buds are ready. Cion-budding (Fig. 115) is a good method.

Bignonia (Trumpet Flower). *Bignoniaceæ.*

Increased by seeds or layering, or, in early spring, by cuttings made from good strong shoots, with two or three joints. Place cuttings of tender sorts in a well-drained pot of sandy soil, under a bell-glass, in bottom heat. Also by seeds. *B. radicans* propagates readily from root cuttings.

Billardiera (Apple Berry). *Pittosporaceæ.*

Increased by seeds, and by cuttings placed in a pot of sandy soil, under a bell-glass, in gentle heat.

Billbergia. *Bromeliaceæ.*

Propagated by suckers, which are taken from the base of the plant after flowering, when they have attained a good size. The best method to adopt is as follows: Hold the sucker in the hand and gently twist it off the stem; next, trim the base by the removal of a few of the lower leaves, and then insert each sucker separately in a small pot, in sharp soil. A bottom heat of about 80° will greatly facilitate new root growth; failing this, they will root freely in the temperature of a stove, if placed in a shaded position for two or three weeks, after which they will bear increased light and sunshine during the latter part of the day.

Billberry. See Vaccinium.

Biota. See Thuya.

Birch. See Betula.

Birthwort. See Aristolochia.

Bitter Sweet. See Celastrus and Solanum.

Blackberry (*Rubus villosus* and vars.). *Rosaceæ.*

New varieties are obtained from seeds, which may be sown as soon as they are cleaned from the ripe fruit, or which may be stratified until the next spring. If the soil is in prime condition, fall sowing is preferable.

Varieties are multiplied by suckers and by root cuttings. The suckers spring up freely about the old plants, especially if the roots are broken by the cultivator; but they have few fibrous roots, and are inferior. The best plants are obtained from root cuttings (Fig. 62). Roots from one-fourth to three-eighths inch in diameter are selected for this purpose. The roots are dug in the fall, cut into pieces an inch or two long, and stored until early spring. They may be buried in boxes of sand after the manner of stratified seeds, or stored in a cool cellar; callusing proceeds most rapidly in a cellar. The pieces are planted horizontally an inch or two deep, in loose, rich soil. It is best to put them in a frame and give them slight bottom heat, although they will grow if planted in the open in April or May, but the plants will make much less growth the first season. Some varieties do not strike quickly without bottom heat. When the variety is scarce, shorter and slenderer pieces of root may be used, but these demand bottom heat. The heat in the frames is usually supplied by manure, or the heat of the sun under the glass may be sufficient. In these frames the cuttings may be started in the north late in March, or some six or eight weeks before the plants can be set out-doors without protection. When the weather has become somewhat settled, the plants may be planted out, and by fall they will be 2 to 3 feet high. See Dewberry.

Bladder-nut. See Staphylea.

Bladder Senna. See Colutea.

Blandfordia. *Liliaceæ.*

Propagated by seeds and offsets, or by division of the old plants, which must be done when repotting.

Blazing Star. See Liatris.

Bleeding Heart. See Dicentra.

Bletia. *Orchidaceæ.*

Propagated by divisions, which should be made after the plants have finished flowering, or previous to their starting into growth. These are terrestrial, and their flat, roundish pseudo-bulbs are usually under ground. They bear division well, especially *B. hyacinthina*, which may be cut up into pieces consisting of a single pseudo-bulb. (See under Orchids.)

Blood Flower. See Hæmanthus.

Blood-root. See Sanguinaria and Hæmodorum.

Blueberry. See Vaccinium.

Blue-eyed Grass. See Sisyrinchium.

Blumenbachia. *Loasaceæ.*

Propagated by seeds sown in pots in spring, and placed in a gentle heat.

Bocconia. *Papaveraceæ.*

Some species grow well from seed. By young suckers, taken from established plants during summer. Cuttings taken from the axils of the large leaves during early summer push freely, so that they will have plenty of roots before winter sets in. Root cuttings of *B. cordata* strike freely.

Boltonia. *Compositæ.*

Increased by divisions of the root in spring. Seeds.

Bomarea. *Amaryllidaceæ.*

Propagated by seeds, which may be sown in a warm house. Also increased by careful division of the underground stem. In making a division, it is necessary to observe that the part taken has some roots by which to live till new ones are formed.

Bombax (Silk Cotton Tree). *Malvaceæ.*

Plants raised from seeds brought from their native habitats make the best trees. Increased by cuttings, which will root readily if not too ripe. They should be taken off at a joint, and placed in sand under a bell-glass, in moist heat.

Borago. *Borraginaceæ.*

Propagated by seeds sown from spring to autumn in any good garden soil. Also by divisions in spring, or by striking cuttings in a coldframe.

Borassus. *Palmaceæ.*

Increased by seeds sown in a strong bottom heat.

Boronia. *Rutaceæ.*

Increased by seed. By young cuttings, or those made from half-ripened wood. Place these in a thoroughly drained pot of sandy soil, with one inch of sand on the surface, and cover with a bell-glass.

Borreria. *Rubiaceæ.*

Propagated by cuttings. Those of the perennial kinds strike root readily in a light soil, in heat. The annual kinds require a similar treatment to other tender annuals.

Boston Ivy, and **Boston Vine.** See Ampelopsis and Myrsiphyllum.

Boswellia (Olibanum Tree). *Burseraceæ.*

Increased easily by cuttings in sand under a glass.

Botrychium. See Ferns.

Bouchea. *Verbenaceæ.*

Increased during spring by cuttings, placed in sand under a glass and in a gentle heat.

Bougainvillea. *Nyctaginaceæ.*

Propagated by cuttings from the half-ripened wood. Place in sandy soil, in a brisk heat. Also grown from root cuttings.

Bouncing Bet. See Saponaria.

Boussingaultia (Madeira Vine). *Chenopodiaceæ.*

Increased by seeds, and easily by means of the tubercles of the stem. Also by the tubers.

Bouvardia. *Rubiaceæ.*

Generally propagated by root cuttings, which strike readily. Cuttings of shoots will also grow, if struck in heat.

Bowiea. *Liliaceæ.*

Propagated by seeds or offsets.

Box Elder. See Negundo.

Box Thorn. See Lycium.

Box Tree. See Buxus.

Brachycome (Swan River Daisy). *Compositæ.*

Propagated by seeds sown in early spring, in a gentle hotbed, or they may be sown thinly outdoors, late in spring.

Brahea. *Palmaceæ.*

Propagated by seeds in heat.

Brassia. *Orchidaceæ.*

Increased by dividing the plant when growth has commenced. (See under Orchids.)

Bravoa. *Amaryllidaceæ.*

Propagated by seeds sown as soon as ripe, and by offsets in autumn.

Bread Fruit. See Artocarpus.

Bread Nut. See Brosimum.

Bredia. *Melastomaceæ.*

Increased by seeds, and by cuttings from the ripened shoots placed in sandy loam, under a hand-glass, in heat.

Briza (Quaking Grass). *Gramineæ.*

Propagated by seeds, which may be sown in spring or in autumn.

Broccoli. See Cabbage.

Brodiæa. *Liliaceæ.*

Increased by offsets, which should be left undisturbed with the parent bulbs till they reach a flowering state, when they may be divided and planted in autumn.

Bromelia. *Bromeliaceæ.*

Some are propagated by seeds. All by cuttings inserted in sand, in heat.

Bromus. *Gramineæ.*

Increased by seeds sown outside in late summer or in spring, thinning out when necessary.

Brongniartia. *Leguminosæ.*

Increased by cuttings of the young shoots, which, if firm at the base, will root in sand under a bell-glass, in a cool house.

Brosimum (Bread Nut). *Urticaceæ.*

Propagated by cuttings of ripe wood with their leaves on. Place in sand in moist heat.

Broughtonia. *Orchidaceæ.*

Increased by dividing the plant. (See under Orchids.)

Broussonetia (Paper Mulberry). *Urticaceæ.*

Propagated by seeds, sown when ripe or kept till the following spring; and by suckers and cuttings of ripened wood, in a cool house.

Browallia. *Scrophulariaceæ.*

Seeds. To have blooming plants for the holidays, they are propagated by seeds sown in late summer in pans or pots of light, rich, sandy soil, and kept in a close frame or hand-light, where they can be shaded till germination takes place.

Brownea. *Leguminosæ.*

Increased by cuttings from the ripened wood; place in sand under a hand-glass, in moist heat.

Brucea. *Simarubaceæ.*

Increased by cuttings from ripened wood, which will root freely in a pot of sand under a hand-glass, in moderate heat.

Brugmansia. See Datura.

Brunfelsia, Franciscea. *Scrophulariaceæ.*

Propagated by cuttings placed in sand under a bell-glass in moderate heat. When rooted, place in pots with a compost of loam, leaf-soil, peat and sand.

Brunsvigia. *Amaryllidaceæ.*

Increased by offsets of considerable size. They should be potted carefully in a mixture of sandy loam and peat, with good drainage, and kept tolerably warm and close until established; water sparingly until root action has commenced. The best place for growing the offsets into a flowering size is on a shelf near the glass, in a temperature of from 50° to 55°.

Brussels Sprouts. See Cabbage.

Bryonia. *Cucurbitaceæ.*

Propagated by seeds, or by divisions of the tuber. Cuttings of the shoots will also strike (but with difficulty) in water.

Bryophyllum. *Crassulaceæ.*

Propagated by cuttings; or by simply laying the leaf on moist sand or moss, and at the indentations upon the margin plantlets will appear. (See Fig. 77.)

Buceras. See Terminalia.

Buckbean. See Menyanthes.

Buckeye. See Æsculus.

Bucklandia. *Hamamelideæ.*

Increased by cuttings of ripened shoots placed in sandy loam under a hand-glass, in moderate heat. Water carefully, for they are liable to rot off.

Buckthorn. See Rhamnus.

Buckwheat (*Fagopyrum esculentum* and *F. Tataricum*). *Polygonaceæ.*

Propagated by seeds.

Buffalo Berry. See Shepherdia.

Bugwort. See Cimicifuga.

Bulbine. *Liliaceæ.*

The bulbous rooted species by offsets, and the herbaceous sorts by suckers and divisions. Also by cuttings.

Bulbocodium. *Liliaceæ.*

Increased by offsets in a rich, sandy loam. Take up the bulbs, divide and replant them every second year, handling in autumn and renewing the soil or planting in new positions.

Bulbophyllum, Anisopetalum. *Orchidaceæ.*

Propagated by division of the pseudo-bulbs.

Bullrush. See Typha and Juncus.

Bupleurum (Hare's Ear). *Umbelliferæ.*

The annuals by seeds sown in spring outdoors; the herbaceous perennials may be increased by divisions made in autumn or spring, and the greenhouse species by cuttings made in spring.

Burchardia. *Liliaceæ.*

Propagated by offsets or divisions made just previous to potting in spring. It is best to repot annually. Good drainage should be allowed, and the plant must not be potted too firmly.

Burchellia. *Rubiaceæ.*

Increased by cuttings, not too ripe, planted in sand and placed under a hand-glass, in a gentle heat.

Burlingtonia. *Orchidaceæ.*

Increased by dividing the plant. (See under Orchids.)

Burnet, or **Poterium** (*Sanguisorba*). *Rosaceæ.*

Propagated by seeds and division.

Burning Bush. See Euonymus.

Bursera. *Burseraceæ.*

Propagated by cuttings placed under a bell-glass, with bottom heat.

Butcher's Broom. See Ruscus.

Butomus (Flowering Rush). *Alismaceæ.*

Increased by seeds, or by divisions of the roots in spring.

Buttercup. See Ranunculus.

Butternut. See Juglans.

Butterwort. See Pinguicula.

Buttonwood. See Platanus.

Buxus (Box). *Euphorbiaceæ.*

Propagated by seeds sown as soon as ripe, in any light, well-drained soil. They can be increased by suckers and divisions; by layers of young or old wood, made in autumn or early spring; by cuttings made of the young shoots, from 4 to 6 inches in length, in a sandy place in spring or fall. The latter method is the better way in this country, and in the north the cuttings should be handled under glass.

Byrsonima. *Malpighiaceæ.*

Increased by cuttings of half-ripened shoots in sand under a hand-glass, in moist bottom heat.

Cabbage (*Brassica oleracea*, and vars.). *Cruciferæ.*

Seeds. They may be sown in the open ground in spring, or in the fall and the young plants wintered in a coldframe, or in a hotbed or forcing house in late winter or spring. Brussels sprouts, broccoli and cauliflower are treated in the same manner.

Cabbage Palm. See Areca.

Cabomba. *Nymphæaceæ.*

Propagated by root divisions; also seeds. See Nymphæa.

Cacalia. See Senecio.

Cacao. See Theobroma.

Cactus. *Cactaceæ.*

Propagation by seeds is not often adopted, as it is a very slow method. The seeds should be sown in very sandy soil, and placed in a semi-shady position until germination commences, when they may be exposed and very carefully watered. Usually propagated by cuttings or offsets, which should be made with a sharp cut, and laid upon a sunny shelf or on dry sand until the wound is healed and roots emitted, when they should be potted in sandy soil. Place in a bench and keep syringed. Some of the less fleshy types may not require this preliminary "curing" or drying. A cereus cutting is shown in Fig. 72. (For an elaborate account of the propagation of cacti by cuttings, see Arloing, Ann. des Sci. Nat. 6th Ser. iv. pp. 5 to 61, with plates, 1876.) Grafting is resorted to with weak kinds, which will not grow freely except upon the stock of a stronger species; and by this means, also, such kinds can be kept from the damp soil, which frequently causes decay. The stocks usually employed are those of *Cereus tortuosus*, *C. Peruvianus*, *Pereskia aculeata*, etc., according to the species intended for working; they readily unite with each other. If the cion and stock are both slender, cleft-grafting should be adopted; if both are broad it is best to make horizontal sections, placing them together and securing in proper position by tying with raffia, but not too tightly, or the surface may be injured. See Fig. 134.

Cæsalpinia. *Leguminosæ.*

Increased by cuttings, which are somewhat difficult to root but may succeed if taken from the plant in a growing state and planted in sand with a hand-glass over them in heat.

Cajanus. *Leguminosæ.*

Plants are usually raised from seeds obtained from the West Indian Islands and India. Also grown from young cuttings, put in sand with a hand-glass over them, in heat.

Caladium. *Aroideæ.*

Increased by tubers, which have been kept dry or rested for some time. Place in small pots in a stove or pit, where the night temperature is maintained at from 60° to 65°, and syringed daily once or twice at least. Large tubers, if sound, may be divided and the pieces potted. Some also by cuttings. Taro is the tuberous roots of *C. esculentum.*

Calamagrostis. *Gramineæ.*

Increased by seeds sown in autumn or spring. *C. arenaria* (now *Ammophila arundinacea*), used for holding sands along seashores, is propagated by division, and can probably be handled easily by root cuttings.

Calamintha. *Labiatæ.*

Increased by seeds, root divisions, or cuttings in spring.

Calamus. *Palmaceæ.*

Increased by seeds.

Calandrinia. *Portulacaceæ.*

Increased by seeds sown in pots where they are intended to flower, as transplantation, unless performed with more than ordinary care, will check their growth or result in loss.

Calanthe. *Orchidaceæ.*

As a rule, the natural annual increase in the number of pseudo-bulbs meets the requirements of most cultivators.

Where a quick propagation is desired, it may be performed by dividing the pseudo-bulbs transversely; after allowing the raw surface to callus, the upper part should be set on moist sand, and several buds will form around the base. The bottom portion may be used in the ordinary way. Another plan is to divide the pseudo-bulbs lengthwise into two or more pieces. (See under Orchids.)

Calathea, or **Maranta.** *Scitamineæ.*

Increased by division in summer or any time between that and the spring months. When making divisions, see that each crown is well furnished with roots.

Calceolaria (Slipperwort). *Scrophulariaceæ.*

Herbaceous kinds increased by seeds sown from June to August on pans of light, sandy soil, which should be soaked with water before sowing. Care must be taken to make the surface of the soil level, and also to sow the seeds as evenly as possible. It is better not to cover with soil, but a sheet of glass should be laid over the pan, which must be placed in a shady part of the greenhouse or coldframe until the young plants show the first leaf. The glass can then be gradually removed. The shrubby kinds, by seeds and by cuttings in August. Place in a coldframe or bench facing the north, in sandy soil, and, when rooted, pot off into 3-inch pots.

Calendula (Pot-Marigold). *Compositæ.*

Increased by seeds; also by cuttings, which thrive well in a compost of loam and peat.

Caliphruria. *Amaryllidaceæ.*

Propagated by bulbels. After flowering, the plants should have a slight heat, and when starting into new growth should be repotted.

Calla. See Richardia.

Calliandra. *Leguminosæ.*

Increased by cuttings of rather firm young wood, in sand under a hand-glass, in heat.

Callicarpa (French Mulberry). *Verbenaceæ.*

Propagated by seeds, divisions, or by cuttings of the young shoots, the last with the same treatment as fuchsia.

Calliopsis. See Coreopsis.

Calliprora. *Liliaceæ.*

Propagated by offsets, which should not be removed from the parent bulbs until they are of good size.

Callipsyche. *Amaryllidaceæ.*

Propagated by seeds and bulbels.

Callirrhöe (Poppy-Mallow). *Malvaceæ.*

Perennials by seeds, divisions of roots, and cuttings; the annuals by seeds only. Cuttings should be started in sandy soil in a frame.

Callistachys. See Oxylobium.

Callistemon. *Myrtaceæ.*

Increased by seeds, and by ripened cuttings in sand under a glass.

Callistephus, Callistemma (China Aster). *Compositæ.*

Propagated by seeds, which should be sown under cover in spring, or seeds for late plants may be sown in the open.

Callitris, Frenela. *Coniferæ.*

Increased by seeds, or by cuttings inserted under a hand-light in autumn, and wintered in a cold pit.

Calluna (Heather). *Ericaceæ.*

Propagated by cuttings of the tender shoots inserted in pure sand under glass in a cool house in autumn.

Calochortus (Mariposa Lily). *Liliaceæ.*

Propagated by seeds, offsets, and by the tiny bulblets on the upper portion of the stem. Sow seeds as soon as ripe, or early in the year, thinly in pans, so that the young plants may pass a second season in the seed-pots or pans. Place in a cool house or frame, and keep the plants close to the glass during their early stages, as they are very liable to damp-off. Early the third season pot off and plant singly, encouraging them to grow freely. The offsets are best removed when the plants are in a dormant state, placed in pots or pans, or planted out in pits or frames until they reach flowering size.

Calodendron. *Rutaceæ.*

Increased by cuttings of half-ripened wood placed in sand under a glass, in gentle bottom heat.

Calophyllum. *Guttiferæ.*

Increased by cuttings made from the half-ripened shoots, which root freely in sand if placed under a glass in bottom heat.

Calopogon. *Orchidaceæ.*

Increased by offsets taken from the tuberous roots.

Calothamnus. *Myrtaceæ.*

Increased by cuttings of young wood, firm at the base. Place in sand and cover with a hand-glass.

Caltha (Marsh Marigold, "Cowslip" in America). *Ranunculaceæ.*

Propagated by seeds sown as soon as ripe, or by dividing the roots in early spring, or in summer after flowering.

Calycanthus (Sweet-scented Shrub, Allspice). *Calycanthaceæ.*

Increased by seeds sown in a coldframe; by divisions or offsets, and by layers put down in summer.

Calypso. *Orchidaceæ.*

Increased by offsets.

Calystegia (Hedge Bindweed, Bearbind). *Convolvulaceæ.*

Propagated by seeds sown in spring, or by dividing the plants.

Camassia. *Liliaceæ.*

Propagated by seeds sown in a warm situation outdoors or in pots or boxes under glass. The young plants should remain at least two years in the seed-beds. Also increased by offsets, which are produced very freely, and which should be removed either when in a dormant condition, or just previous to starting into fresh growth, and arranged in clumps or rows, placing a little sand about them.

Camellia, including Thea ("Japonica," Japanese Rose). *Ternstrœmiaceæ.*

The single red camellia by either seeds, layers or cuttings. Double and variegated camellias by layers, but cuttings will succeed. Seeds give suitable stocks on which to inarch or graft the rarer kinds. The ripened shoots of the preceding summer should be taken off in August. Two or three of the lower leaves should be removed, and the cuttings planted firmly in the soil with a dibble. The pans containing the cuttings should be kept in a box or coldframe, without being covered with glass, but snaded during bright sunshine. In the following spring, such as have struck will begin to push, when they need to be placed in a gentle heat. Make cuttings during winter while one-year-old wood is dormant. Inarching or grafting is done in early spring, as soon as growth commences (Figs. 138, 139).

Camomile. See Anthemis.

Campanula (Bell-flower, Slipperwort). *Campanulaceæ.*

Increased by seeds. The perennials are also propagated by dividing the roots, or by young cuttings in spring.

Camphora (Camphor-tree). *Lauraceæ.*

Increased by cuttings and seeds.

Campion. See Silene.

Candollea. *Dilleniaceæ.*

Increased sometimes by seeds, but usually by cuttings, which will root if placed under a hand-glass in a compost of equal parts loam and peat, with enough sand to render the whole porous.

Candytuft. See Iberis.

Canella. *Canellaceæ.*

Increased by well-ripened cuttings taken off at the joint. They will root in sand under a hand-glass, with bottom heat, in spring; but care should be taken not to deprive them of any of their leaves.

Canna (Indian Shot). *Scitamineæ.*

Propagated by seeds sown in heat in late winter. The seeds are very hard, and germination will be materially stimulated if they are filed (see page 18) and then soaked in tepid water for twenty-four hours. They should be sown thinly in pans (a mixture of sand and leaf-loam is best for them), and a covering of one and one-half or two inches of earth is not excessive. It is a good plan to sow the seeds singly in small pots. Seeds give new varieties. Also increased by divisions; they form a large crown or stool of strong buds, each portion of which, with bud and roots attached, may be converted into an independent plant (Fig. 27). Named varieties are multiplied in this manner. For ordinary planting-out, the divided crowns are usually set directly in the open. If very early effects are desired, however, the pieces may be started or in pots; and this is always done by dealers, for they send out growing plants. As soon as frost comes in the fall, the tops are cut, and the crowns lifted and stored in a dry, cool cellar on shelves. Care must be taken that the cellar be given plenty of air until the roots are thoroughly cured, else they may rot.

Cannabis (Hemp). *Urticaceæ.*

Propagated by seeds sown in spring.

Cantua. *Polemoniaceæ.*

Increased by cuttings placed in sand under glass.

Capparis (Caper). *Capparidaceæ.*

Propagated by cuttings of ripe shoots, which will root in sand under glass, in moist heat. Seeds, when obtainable.

Capsicum. See Pepper, Red.

Caragana (Siberian Pea-tree). *Leguminosæ.*

Propagated by seeds and by root cuttings; the low-growing shrubs by seeds and layers. Caraganas are generally increased by grafting on *C. arborescens*, which is easily raised from seeds, sown when ripe or in spring.

Cardamine (Lady's Smock). *Cruciferæ.*

Seeds. Propagated easily by division after flowering.

Cardinal-flower. See Lobelia.

Carex (Sedge). *Cyperaceæ.*

Propagated by seeds, or by division, usually the latter. Seeds often lie dormant the first year.

Careya. *Myrtaceæ.*

Propagated by division, or by ripened cuttings, which root freely if planted in sand under a hand-glass, and placed in moist bottom heat.

Carica (Papaw-tree). *Passifloraceæ.*

Propagated by cuttings of ripe shoots with their leaves on. They root readily in a sandy soil and in a gentle bottom heat. Seeds, when obtainable. Sow in heat.

Carnation. *Caryophyllaceæ.*

By propagating by seeds, new varieties are raised. Sow the seeds in spring, and in a slight hotbed or in a greenhouse. Also propagated by layering, which should be done at the end of July or the beginning of August. The shoots selected should be denuded of a few of their leaves at the base of the young wood, and a slit must be made from this point upwards, extending through a joint of the bare stem, so that a tongue is formed. This is the method employed in Europe. See Fig. 31.

In this country, always increased by cuttings. It is necessary to have a slight bottom heat, and on it put four or five inches of light soil, covered with clean sand. The cuttings must be long enough to have a tolerably firm base, and they should either be taken with a heel or cut off at a joint, and firmly inserted in the soil. See Fig. 69, *b*.

Carob, Algaroba, or St. John's Bread (*Ceratonia siliqua*). *Leguminosæ*.

Stocks are obtained by seeds. The seeds are often treated to scalding water before sowing, in the same manner as locust seeds. Varieties are grafted or budded on the seedlings, or they may be multiplied by means of hard-wood cuttings in frames.

Carpinus (Hornbeam). *Cupuliferæ*.

Increased by seeds, which germinate irregularly. Varieties propagated by budding or grafting on seedling stocks.

Carrion Flower. See Stapelia.

Carthamus (Safflower). *Compositæ*.

Increased by seeds sown in a gentle heat in spring.

Carya. See Hicoria.

Caryocar (Butternut). *Ternstrœmiaceæ*.

Increased by ripened cuttings, which will root in sand in heat. Seeds, if obtainable.

Caryophyllus (Clove-tree). *Myrtaceæ*.

Increased by cuttings of firm shoots with the leaves left on. These will root if planted in sand in a moist heat.

Caryopteris. *Verbenaceæ*.

Propagated by seeds, by division, or by cuttings.

Caryota. *Palmaceæ*.

Increased easily by seeds or by suckers.

Cashew. See Anacardium.

Cassandra (Leather Leaf). *Ericaceæ.*

Propagated by seeds very carefully sown, or by layers. Sow seeds in peat or on live sphagnum moss.

Cassava (*Manihot Aipe*). *Euphorbiaceæ.*

Propagated by cuttings of the stem and by suckers. Cut the large main stalks into pieces from 4 to 6 inches long, and set them perpendicularly into the ground in the field. The cuttings can be struck at various times, but spring is usually preferred. The stalks can be kept over winter by covering with sand on a dry knoll, placing the stalks and sand in layers. Cover the whole with boards to shed the water. Suckers which appear during summer can be removed and planted or made into cuttings.

Cassia. *Leguminosæ.*

Annuals and biennials by seeds, which must be sown in spring, in a gentle heat. The shrubby species by cuttings of half-ripened shoots, which will root in heat. *C. Marylandica* also by division.

Cassine. *Celastraceæ.*

Increased by ripened cuttings, which will readily strike root if planted in a pot of sand with glass over them.

Castalia. See Nymphæa.

Castanea. See Chestnut and Chinquapin.

Castor Bean. See Ricinus.

Casuarina (Beefwood). *Casuarineæ.*

Propagated by seeds; or by cuttings made of half-ripened shoots, placed in sand under glass.

Catalpa. *Bignoniaceæ.*

Increased by seeds, and by cuttings made of the ripe wood. The named varieties and *C. Bungei* are propagated by soft cuttings in June and July. Grafts are also used, setting them upon seedlings of *C. speciosa* or *C. bignonioides.*

Catananche. *Compositæ.*

Increased by seeds, which should be sown in spring. Also by division.

Catchfly. See Silene.

Catesbæa (Lily Thorn). *Rubiaceæ.*

Propagated by cuttings planted in sand in spring, and plunged in heat.

Catnip, or **Catmint** (*Nepeta Cataria*). *Labiatæ.*

Seeds. Division.

Cat-Tail. See Typha.

Cattleya. *Orchidaceæ.*

Increased by the pseudo-bulbs. (See under Orchids.)

Cauliflower. See Cabbage.

Caulophyllum (Blue Cohosh). *Berberidaceæ.*

Propagated by divisions of the roots, made in early spring or after flowering. Also by seeds, stratified.

Ceanothus. *Rhamnaceæ.*

Increased by layers, which is the readiest way of obtaining strong plants, or by cuttings, which should be inserted in a coldframe. Stratified seeds.

Cedar. See Cedrus and Juniperus.

Cedrela (Bastard Cedar). *Meliaceæ.*

Increased by large ripened cuttings, placed in sand, in heat. *C. Sinensis* by root-cuttings.

Cedronella. *Labiatæ.*

The herbaceous species by division of the roots or by cuttings of young wood. *C. triphylla* by cuttings.

Cedrus (Cedar). *Coniferæ.*

Increased by seeds, which are difficult to extract from the cones. Gather the cones in spring, and sow the seeds immediately in pans. Varieties are propagated by veneer grafts.

Celastrus (Staff-tree, Bitter-sweet). *Celastraceæ.*

Propagated by seeds and suckers; also by layering the hardy species in autumn. Ripened cuttings will root freely in a compost of loam, peat and sand.

Celery (*Apium graveolens*). *Umbelliferæ.*

By seeds, as described on pages 5, 22; or, for the early crop, sow under glass, as in a hotbed.

Celosia (Cockscomb). *Amarantaceæ.*

Propagated by seed sown in spring, in pans or frames, or in the open.

Celsia. *Scrophulariaceæ.*

Increased by seeds, which may be sown in the open border and thinned out for flowering, or raised in nursery beds and transplanted. *C. Arcturus* should be increased by cuttings, the young wood striking freely in a cool house or frame.

Celtis (Nettle-tree). *Urticaceæ.*

Increased by seeds, which should be sown as soon as ripe. By layers, and by cuttings of ripened shoots in autumn.

Centaurea. *Compositæ.*

Annuals by seeds, which may be sown in the open border. To propagate *C. Cineraria* and some others, sow seeds in August in slight heat, or make cuttings about the beginning of September.

Centranthus. *Valerianaceæ.*

Increased by seeds sown in spring.

Centropogon. *Lobeliaceæ.*

Increased by seeds, by divisions and by cuttings from any young shoots 3 or 4 inches long. Take off with a heel and place in sharp sandy soil, close around the edge of the pot, and then keep close under a propagating box, in a temperature ranging between 60° and °70.

Cephalanthus (Button-Bush). *Rubiaceæ.*

Seeds. Propagated by layers, or ripened cuttings in autumn.

Cerastium. *Caryophyllaceæ.*

Propagated by seeds and divisions, or by cuttings inserted in the open ground in a shady place, after flowering.

Ceratiola. *Empetraceæ.*

Increased by seeds and by cuttings, which should be placed in sandy soil under glass.

Ceratonia. See Carob.

Ceratozamia. *Cycadaceæ.*

By seeds, and sometimes by suckers and divisions, but imported plants give most satisfaction. See Cycas.

Cercidiphyllum. *Magnoliaceæ.*

Propagated by tender cuttings made during the summer, and slightly wilted before placing in the frames. By seeds, when procurable.

Cercis (Red-bud, Judas-tree). *Leguminosæ.*

Propagated by seeds, sown about the end of March on a bed of light soil, in a gentle heat. They may also be increased by layers, but plants raised from seeds thrive best. It is not necessary to stratify the seeds. *C. Japonica* is grown from soft cuttings in early summer.

Cereus. See Cactus.

Ceropegia. *Asclepiadaceæ.*

Propagated by cuttings of small side shoots made in spring, which will root in sand, in heat, with or without a glass covering.

Cestrum, including Habrothamnus. *Solanaceæ.*

Propagated by cuttings in August, or whenever the wood is fit.

Chamæcyparis. *Coniferæ.*

Propagated by seeds freely, also by layers, but mainly by cuttings put in during October in a cool greenhouse. Select young side shoots with a heel; insert in well-drained pots of sandy soil, and place in a close coldframe, keeping fairly moist through the winter. In February they should be callused, and should be placed in gentle heat, where they will root freely. See Retinospora.

Chamærops, including Corypha. *Palmaceæ.*

Increased by seeds, or by suckers, which generally appear in considerable quantities.

Chamomile. See Anthemis.

Chard. See Beet.

Cheilanthes. See Ferns.

Chelone (Turtle-head). *Scrophulariaceæ.*

Increased by means of seeds. Also by dividing the plant during fall. Young cuttings inserted in sandy soil in a coldframe grow well.

Cherry (*Prunus Avium*, *P. Cerasus*, etc.). *Rosaceæ.*

Cherry stocks are commonly grown from seeds. If the ground is in readiness, and is in proper condition, the seeds may be planted in fall, or even as soon as they are ripe. If stored until spring, they must be stratified and kept very cool to prevent germination, and they should be sown at the earliest possible moment. They do not need to be cracked by hand. Care must be taken that cherry pits do not become hard and dry. This precaution is more important with cherries than with peaches and plums. At the close of the first season, the seedlings will be a foot or foot and a-half high, large enough to transplant into nursery rows, after the manner of apples, where they are budded the following season. In warm climates the pits are sometimes cracked as soon as they are gathered, and the "meats" planted immediately. They will then make stocks fit for grafting the following winter, or for transplanting and budding the following summer. Cherry seeds must never be allowed to become so dry that the meat is hard and brittle.

Cherries, in common with other stone fruits, grow readily from root-cuttings, in the same manner as blackberries. They do better if started over a gentle heat.

The Mazzard cherry is the stock upon which cherries are recommended to be worked. It is simply a hardy and vigorous variety, with inferior fruit, of the common sweet cherry (*Prunus Avium*). Seeds of this are readily procured in this country. As a matter of fact, however, nearly all sour cherries are worked upon the Mahaleb in this country, as they take better upon it, and the stocks are cheap. Sweet cherries are often budded upon the Mahaleb, but it is a question if such practice is best. The Mazzard is such a strong grower that the bud is often "drowned out" by the flow of sap. In order to avoid this exuberance, nurserymen often pinch in the tips of the stocks a few days before they are to be worked. The Mazzard is also liable to leaf-blight, and to serious injury from the black aphis, so that the bark often sets before the operator has had time to finish his plantation. Mazzards usually have a shorter budding season than Mahalebs, and are less uniform in behavior; and for these reasons, Mahalebs are widely used. This is a distinct species, *Prunus Mahaleb*, from Southern Europe. The seeds or stocks are imported. Màhaleb stocks are recommended in the books for dwarfing the cherry, but the dwarfing depends more upon pruning

Cherry, continued.
than upon the Mahaleb root. The Mahaleb is naturally a smaller tree than the Mazzard, however. It is said that the Mahaleb is better adapted to heavy clay soils than the Mazzard, but in practice it is used indiscriminately for all soils and nearly all varieties.

Morello (*Prunus Cerasus*) stocks will no doubt prove to be valuable in the northwest, where great hardiness is demanded. Seedlings do not sprout or sucker badly, but the natural suckers, which are sometimes used for stocks, are likely to be more troublesome in this respect. If strong-growing tops are worked on Morello stocks, however, there is usually little annoyance from suckering. Mahaleb stocks are generally used for the Morello cherries.

It is probable that some of the native American cherries can be used as stocks. The common wild red, pin, pigeon or bird cherry (*Prunus Pennsylvanica*) has already been used to some extent. The sweet and sour cherries unite readily with it, and bear very early. It is yet to be determined how long the trees will persist, but there are trees known which are sixteen or eighteen years old, and which are still healthy and vigorous. It is considered to be a very promising stock for the cold prairie states. The dwarf or sand cherries (*Prunus pumila* and *P. Besseyi*) give promise as dwarf stocks.

Cherry stocks are worked both by budding and grafting. Budding is the common method. The stocks should be fit to work the season they are transplanted, or in the second summer from seed. Such as are too small for working then may be allowed to stand until the following year; or if the number is small, the poor ones are rooted out.

In the west, where great hardiness is required, the varieties are crown-grafted upon Mazzard stocks in winter. Yearling stocks are used, and the cions are from 6 to 10 inches long. When planted, only the top bud should be left above ground. The cion strikes roots, and own-rooted trees are obtained.

The ornamental cherries are worked upon the same stocks as the fruit-bearing sorts. Mahaleb and Mazzard are commonly used for all species, the latter for weeping forms which need to be worked high.

Cherry trees can be top-grafted as readily as apple or pear trees, and the same methods are employed. They are usually grafted very early in the spring. The chief

Cherry, **concluded.**

requisite is that the cions be completely dormant. They should be cut in winter and stored in an ice-house or a cold cellar.

Chervil (*Chærophyllum bulbosum* and *Scandix cerefolium*). *Umbelliferæ*.

Seeds, sown much the same as celery seeds, but the plants are usually allowed to stand where sown. Seed is often sown in autumn.

Chestnut (*Castanea sativa* and var. *Americana*, and *C. Japonica*). *Cupuliferæ*.

Chestnut stocks are grown from seed. Difficulty is sometimes experienced in keeping the seeds, as they lose their vitality if dried too hard, and are likely to become moldy if allowed to remain moist. The surest way is to allow the nuts to become well dried off or "seasoned" in the fall, and then stratify them in a box with three or four times as much sand as chestnuts, and bury the box a foot or two deep in a warm soil until spring. They do not always keep well if stored or stratified in a cellar. Fall planting exposes the nuts to squirrels and mice. American stocks are better than European, because the latter are tender in the north.

The stocks are worked by whip-grafting above ground, the wound being well tied and protected by waxed cloth. Care should be taken to have the stock and cion about the same size, in order to secure a good union. Chestnuts can be cleft-grafted like apples and pears; but in small trees it is preferable to set the grafts below ground, as in grapes. The cions should be cut early, before they begin to swell, and kept perfectly dormant until the stock begins to push into leaf. Only vigorous stocks should be grafted. The best results are obtained when the stocks have recovered from transplanting, or when they are from three to five years old. The working of chestnut stocks is far from satisfactory in a commercial way. The union is imperfect in many varieties, and usually no more than half the grafts take well and live long. In all nut trees, the skill of the operator is more important than the particular method employed.

Chicory (*Cichorium Intybus*). *Compositæ*.

Seeds, sown in spring where the plants are to grow. Division.

Chilopsis (Desert Willow). *Bignoniaceæ.*

Increased by seeds, or by cuttings of half-ripened shoots in sand under glass, in a gentle bottom heat.

Chimonanthus. *Calycanthaceæ.*

Propagated by layering in the autumn.

China Aster. See Callistephus.

Chinquapin (*Castanea pumila*). *Cupuliferæ.*

By seeds. Can be handled in same manner as chestnut, which see.

Chiococca (Snowberry). *Rubiaceæ.*

Propagated by cuttings, which strike root freely in sand under glass, in heat.

Chionanthus (Fringe-tree). *Oleaceæ.*

Increased by seeds, which should be started in a coldframe. By layers and cuttings. By grafting or budding it on the common ash, it succeeds very well.

Chionodoxa. *Liliaceæ.*

Propagated by seeds, which are produced freely. They should be sown as soon as ripe. By bulbels.

Chironia. *Gentianaceæ.*

Increased by seeds, and by cuttings inserted in sandy soil and placed in a gentle heat in spring.

Chives, or Cives (*Allium Schœnoprasum*). *Liliaceæ.*

Division of the clumps.

Choisya. *Rutaceæ.*

Increased by ripened cuttings.

Christ's Thorn. See Paliurus.

Chrysanthemum. *Compositæ.*

Increased by seeds to obtain new varieties; these should be sown in spring. Division may be made, but this is not often practiced. Usually propagated by cuttings about three inches long, of firm, healthy, short-jointed shoots, which spring from the base of the plant after the flowering season. They should be made in late winter or spring, and placed near the glass of a rather close frame having a temperature of about 45°. If inserted in pots, only the lower leaf should be removed; if in beds, the remaining foliage should also be trimmed to admit air. Insert about half of the cutting, press the soil firmly, and water. Leaf cuttings have been employed. Inarching and grafting may also be performed, when it is desired to grow two or more varieties on one plant.

The time at which chrysanthemum cuttings should be taken depends upon the season at which bloom is wanted, and the methods of cultivation. The plants may be flowered in pots, or in a solid soil bench. Very good small plants may be brought to perfection in 6-inch pots, but the best results, in pot plants, are to be obtained in 8-inch or 10-inch pots. If the plants are to be used for decoration, they should, of course, be grown in pots, but the best results for cut-flowers are usually obtained by growing in the earth. In any case, the cuttings are made from the tips of basal or strong lateral shoots, late in February to May. One form of cutting is shown in Fig. 71. If the plants are to be flowered in pots—in which case they usually mature earlier—the cuttings may be started as late as April, or even June; but if they are grown in the soil and large plants are desired, the cuttings should be taken in February or March. The plants which are flowered in the soil are generally grown in pots until July. The plants are flowered but once, new ones being grown from cuttings each year.

The Marguerite or Paris Daisy (*C. frutescens* and *C. fœniculaceum*) are propagated by cuttings of firm shoots, like geraniums.

Chrysobalanus (Coco Plum). *Rosaceæ.*

Increased by seeds when procurable. Large cuttings, however, taken off at a joint without shortening of leaves, will root readily if planted thinly in a pot of sand, and placed in moist heat with a bell-glass over them.

Chrysocoma (Goldy-locks). *Compositæ.*

Propagated by seeds, or by cuttings of half-ripened shoots, placed in sand under glass. Seeds.

Chrysogonum. *Compositæ.*

Seeds. Increased by dividing the roots in spring.

Chrysophyllum (Star Apple). *Sapotaceæ.*

Increased by seeds when procurable. By cuttings of small, well-ripened shoots, plunged in strong, moist heat.

Cicca (Otaheite Gooseberry). *Euphorbiaceæ.*

Seeds. By cuttings of ripe shoots, which will root in sand if placed under a glass and in bottom heat.

Cimicifuga (Bugwort). *Ranunculaceæ.*

Increased by seeds, sown in a coldframe or border as soon as ripe; or by division of the roots in spring.

Cinchona (Peruvian Bark). *Rubiaceæ.*

Imported seeds, and cuttings taken off when ripe and planted in a pot of sand, under glass, in a moist heat.

Cineraria. *Compositæ.*

Seeds should be sown under glass; those intended for autumn flowering in April and May, those for spring in July and August. Light leaf-mold should be used, and about an equal quantity of fresh sifted loam and sharp sand added, the whole being well mixed. Old cow-manure is a good medium in which to sow (see page 20). Also by divisions and by cuttings.

Cinquefoil. See Potentilla.

Cipura. *Iridaceæ.*

Propagated by seeds, which should be sown in a slight heat in spring; or by bulbels, which are abundantly produced.

Cissampelos. *Menispermaceæ.*

Propagated by cuttings, which root readily in heat.

Cissus. *Vitaceæ.*

Propagated by cuttings in the spring. Choose the weakly shoots that are pruned just before the plants break into new growth, or allow the young shoots to grow to a length of about two inches. Then cut them off, with a small piece of the basal branch adhering to the young wood; or the shoots may be cut off with one or several of these young branchlets on them. Cut the old branch through at the base of each young one, and insert the cutting with this heel of the old wood entire. In this country, usually grown from common green cuttings in summer.

Cistus (Rock Rose). *Cistaceæ.*

Propagated by seeds, by layers or cuttings under frames outside, or inside with a gentle bottom heat; but seedlings always make the best plants. The seeds should be sown early in the spring in pans or boxes in a frame, and lightly covered with sifted sandy mold. Cuttings should be made from 3 to 4 inches long. They may be struck in spring or autumn, in sandy peat, under glass.

Citron (*Citrus Medica*). *Rutaceæ.*

Seeds, which usually reproduce the kind. Mature cuttings, the same as lemon. Also budded on orange, lemon or lime stocks.

Citrus. *Rutaceæ.*

Increased by seeds, layers, cuttings, inarching, grafting and budding. For particular methods, see Ægle, Citron, Kumquat, Lemon, Lime, Orange and Pomelo.

Cladrastis (Yellowwood). *Leguminosæ.*

Propagated by seeds sown in the open air in spring, or by cuttings of the root.

Clarkia. *Onagraceæ.*

Increased by seeds, which may be sown in spring or autumn outdoors.

Clematis (Virgin's Bower). *Ranunculaceæ.*

Clematis may be increased by seeds. The seed-heads should be gathered before autumn, and stratified till the following spring, when the seeds may be sown in light, sandy soil, and placed in gentle heat till they germinate. By layers outside, put in at any time. All the varieties of clematis may also be increased by cuttings made of the young shoots, which may be cut up to every eye and planted in gentle heat. Also by grafting any of the varieties on portions of clematis roots in winter. Good, healthy pieces of root obtained from old plants answer the purpose well. See also Atragene.

Cleome. *Capparidaceæ.*

Increased by seeds sown in a frame in spring, with slight warmth. Ripened cuttings root freely in moderate heat.

Clerodendron, Volkameria. *Verbenaceæ.*

Increased by seed, which, if sown when ripe or in the spring, and grown on in heat, may be converted into flowering plants the second season. Propagated also by cuttings of both green and mature wood; also of roots. Suckers. The climbing varieties do not root quite so readily from cuttings as the others, but cuttings of the ripened wood do well.

Clethra. *Ericaceæ.*

Propagated by seeds (as for Andromeda), divisions and layers. Cuttings taken from the half-ripened wood will root in gentle heat.

Clianthus (Glory Pea, Parrot Beak). *Leguminosæ.*

C. Dampieri is best raised from seeds, which should be sown singly in good-sized pots, when the necessity of first shifting will be obviated. *C. puniceus* and others from cuttings, which strike easily in sand in bottom heat.

Clintonia. *Liliaceæ.*

Propagated by seeds, and by division of the root in spring.

Clitoria. *Leguminosæ.*

The best method of increasing is by seeds. Increased also by cuttings of stubby side shoots, which will root in sandy soil, in heat.

Cliva, Imantophyllum. *Amaryllidaceæ.*

Propagated by seeds or divisions.

Clove-tree. See Caryophyllus.

Clusia (Balsam-tree). *Guttiferæ.*

Increased by cuttings of half-ripened shoots, which will strike in sand, with bottom heat.

Cobæa. *Polemoniaceæ.*

Readily raised from fresh seed in spring, if a gentle bottom heat is supplied. It is often said that the seeds must be placed on edge, but this is a mistake. Exercise care not to keep the seed soil too moist. From cuttings taken when young, in spring, and inserted in pots of sandy soil, placed in gentle bottom heat.

Coccoloba (Seaside Grape). *Polygonaceæ.*

Propagated by seeds and by cuttings of the ripened wood, with leaves entire, and taken off at a joint. These will root freely in sand under glass.

Cocculus, Wendlandia. *Menispermaceæ.*

By seeds. By half-ripened cuttings of side shoots; these will root easily in spring or summer, if planted in sand and placed in bottom heat, under glass.

Cockscomb. See Celosia.

Cocoanut (*Cocos nucifera*). *Palmaceæ.*

The nuts are buried in nursery rows, and the young trees are transplanted. A more common practice is to remove the buried nuts, when they begin to sprout, to the place in which the tree is to stand. A nut is then placed in a hole some two feet deep, which is gradually filled in as the plant grows. In from six to eight years the tree begins to bear. See Palms.

Cocos. *Palmaceæ.*

Most species by seeds in heat. Some by suckers. See Cocoanut, above.

Codiæum, Croton. *Euphorbiaceæ.*

New varieties are produced by seed. Increased by taking off the tops of any strong leading shoots, and making them into cuttings. They may be struck by placing singly in small pots and covering with bell-glasses, in strong, moist heat, where they will soon emit roots, without losing any of the leaves attached at the time they were inserted. Or they may be placed in a bed of sand.

Coffea (Coffee-tree). *Rubiaceæ.*

Propagated by seeds. Also by ripe cuttings, which strike freely in sand under glass, in moist heat; and the young plants so raised produce flowers and fruit more readily than those grown from seed.

Coffee-tree, Kentucky. See Gymnocladus.

Colchicum (Autumn Crocus). *Liliaceæ.*

Seeds, sown as soon as ripe in a protected place. Separation.

Coleus. *Labiatæ.*

Increased by seeds (which grow readily) for new varieties. By cuttings with the greatest freedom at almost any time of the year, and, with a good, moist heat, they will quickly form fine specimens. (Fig. 70.)

Collinsonia. *Labiatæ.*

Increased readily by dividing roots of the perennials in spring; also seeds.

Colocasia. As for Caladium.

Columbine. See Aquilegia.

Colutea (Bladder Senna). *Leguminosæ.*

Propagated by seeds, or by cuttings placed in sandy soil in the autumn.

Combretum. *Combretaceæ.*

Increased by cuttings of side shoots, taken off with a heel, planted in sand under glass, and placed in heat. Seeds, if obtainable.

Comfrey. See Symphytum.

Commelina. *Commelinaceæ.*

Increased by seeds. By cuttings, which will root in sand, in a gentle hotbed.

Comparettia. *Orchidaceæ.*

Increased by division of the plants. (See under Orchids.)

Comptonia (*Myrica asplenifolia*, Sweet Fern). *Myricaceæ.*

Seeds; by dividing the clumps, and by layers, which should be put down in autumn.

Conifers. See the various genera, as Abies, Picea, Larix, Cedrus, Retinospora, Thuya, Juniperus, etc.

Conocarpus (Button-tree). *Combretaceæ.*

Seeds. Increased by cuttings of firm shoots, taken in April, in bottom heat.

Convallaria (Lily-of-the-Valley). *Liliaceæ.*

Increased by "crowns" or "pips" (see Fig. 26), which are the separated growing points of the roots, possessing a strong bud. These crowns can be obtained from any well established bed in the fall, but they are usually imported.

Convolvulus (Bindweed). *Convolvulaceæ.*

Seeds of the hardy annuals should be sown in spring in the open border. The hardy perennials may be increased by seeds sown in spring, by division of the roots, and by young cuttings.

Coptis. *Ranunculaceæ.*

Propagated by seeds and division of the roots.

Cordia. *Borraginaceæ.*

Seeds. Increased by cuttings, green or ripe, which strike root readily in sand, in heat.

Cordyline (Dracæna, Dragon-tree). *Liliaceæ.*

Seeds, if fresh ones are obtainable, for many of the species. The varieties (as the greenhouse dracænas), by cuttings. Chinese layers (Fig. 34) succeed fairly well. The stems of old plants may be cut up in pieces 1 or 2 inches long, and placed at any season in cocoanut fiber or light soil, in the bottom heat of a propagating house. The tops of the plants will also strike as cuttings, and the fleshy base of the stem is sometimes removed and used for propagation. Root cuttings do well in a moderate heat, and are much used. (See Fig. 63.)

Corema (Portugal Crakeberry, Crowberry). *Empetraceæ.*

Seeds.

Coreopsis, Calliopsis. *Compositæ.*

The hardy annuals, which are largely grown under the name of calliopsis for summer ornamentation, by seeds. which should be sown in early spring in a gentle heat, or outside later. The perennials are propagated also by division of the root in autumn or spring, or during the summer by young cuttings, which will strike freely in a coldframe.

Coriander (*Coriandrum sativum*). *Umbelliferæ.*

Seeds sown in fall or spring.

Corn. See Maize.

Corn Salad (Valerianella, several species). *Valerianaceæ.*

Seeds sown in spring, summer or autumn. The plants mature quickly.

Cornus (Dogwood, Osier). *Cornaceæ.*

Increased by seed, suckers of soft wood, layers or cuttings. The herbaceous species, *C. Canadensis* and *C. Suecica*, may be increased by division, as also by seeds. The willow-like cornuses grow from cuttings of ripe wood. *C. stolonifera* and its kin by layers or stolons. Named varieties and some species are budded in many cases, especially all the weak-growing sorts. *Cornus Mas*, raised from seed, is the favorite stock. Shield-budding in late summer and veneer-grafting are most successful. A cutting is shown in Fig. 60.

Coronilla. *Leguminosæ.*

By seeds sown as soon as ripe. The hardy species by

division. Cuttings strike freely if placed in a coldframe or a cool house under a hand-glass in spring, and when callused, introduced to gentle bottom heat.

Cortusa. *Primulaceæ.*

Increased by seed sown as soon as ripe, in a coldframe; also by carefully dividing the roots.

Corydalis. *Fumariaceæ.*

Increased by seeds, or by dividing the plants directly after flowering. The bulbous-rooted species by offsets.

Corylus (Hazel, Filbert, Cob-nut). *Cupuliferæ.*

Propagated by seeds, suckers, layers or cuttings. Grafting and budding are each practicable, and are adopted when growing tall standards or scarce varieties. The seed of all should be sown as soon as gathered, or stored in sand till the following spring. All superior varieties should be increased by suckers or layers. Stools kept for layering must be allowed to make more growth than those used for suckers. Free growth must be encouraged for a year or more, and, any suitable time in winter, the shoots should be bent to the ground, pegged firmly, and covered to the depth of 3 inches with earth. They will be well rooted by the following autumn, and may then be removed and planted out permanently.

Cosmos. *Compositæ.*

Seeds, usually started under glass. The tuberiferous species like Dahlia, which see.

Costus. *Scitamineæ.*

Increased by dividing the roots.

Cotoneaster. *Rosaceæ.*

Propagated readily by seed, which should be sown in spring; by layers or cuttings in autumn, or by grafting on *C. vulgaris*, the common quince, or the hawthorn.

Cotton (Gossypium). *Malvaceæ.*

Seeds commonly. When grown as a curiosity under glass, it may be increased by soft cuttings.

Cotyledon (Navelwort). *Crassulaceæ.*

Increased by seed, offsets, cuttings of the stem, and by leaves. The leaves should be pulled off in autumn, laid on dry sand in pans on a shelf in a propagating or other warm house, and not watered until small plants appear at the ends of the leaves.

Cow-pea. See Vigna.

Cowslip. See Primula and Caltha.

Crambe. *Cruciferæ.*

Increased by seeds, by dividing the roots and by root cuttings. See Sea-kale.

Cranberry (*Vaccinium macrocarpon*). *Ericaceæ.*

The cultivated cranberry is propagated entirely by cuttings. These are made from vigorous young runners, from 6 to 10 inches in length, and they are thrust obliquely into the soil until only an inch or two of the tip projects. Some blunt instrument, as a stick, is commonly used to force them into the sand of cranberry bogs. Planting is done in the spring, and the cuttings are taken just previous to the operation. If cranberry seedlings are desired, the seeds should be sown in flats of peaty earth, which are stored until spring in some protected place, in the manner of stratification boxes. The seeds should be covered lightly, preferably with fine moss. The plants are allowed to grow the first year in the box.

Crassula. *Crassulaceæ.*

Seeds; also by cuttings, which should be taken off and laid for two or three days in the sun to dry before planting.

Cratægus (Haw, Hawthorn). *Rosaceæ.*

Propagated by stratified seeds, which remain dormant for one or two years. Some growers spread the haws in shallow piles in the fall, and allow them to decay, so that most of the pulp is removed before they are stratified. Haws often come irregularly, even from stratified seeds. The varieties are grafted, rarely budded, on common stocks.

Cress (*Lepidium sativum*). *Cruciferæ.*

Seeds, sown at any time of year. See Water Cress.

Cress, American. See Barbarea.

Cress, Rock. See Arabis.

Crinum. *Amaryllidaceæ.*

Increased by seeds, sown singly as soon as ripe in three or four-inch pots, in sandy loam and leaf-mold. Place in a temperature of from 70° to 80°, and keep rather dry until the plants appear, when more moisture should be applied. Also increased by offsets, which should be removed when rather small and potted separately, and grown as recommended for seedlings.

Crithmum. *Umbelliferæ.*

Propagated by seeds sown as soon as ripe, and by divisions.

Crocosmia. *Iridaceæ.*

Propagated by seeds sown in pans in a cold house as soon as possible after maturity. Also by offsets.

Crocus. *Iridaceæ.*

Propagated by seed, sown as soon as ripe or early in spring, the choicer strains in pots or boxes, using a light, sandy soil, and afterwards placing them in a cold pit or frame; the more common varieties may be placed in a warm position outside in a seed-bed. Sow thinly, so that the plants may grow two years in the seed-pan or bed without lifting. By the corms. These may be lifted and replanted, allowing each in its turn to develop new corms below. The following year new corms, or cormels, are also formed by the side of the old corms. These old corms die away annually. Some species increase much more rapidly than others. (See page 31.)

Crotalaria (Rattle-box). *Leguminosæ.*

Increased by seeds. The shrubby kinds by young cuttings, which root freely in sand, under glass, in a cool house.

Croton. See Codiæum.

Crowfoot. See Ranunculus.

Crucianella (Crosswort). *Rubiaceæ.*

Propagated by seeds, by divisions during spring or autumn, and by cuttings.

Cryptomeria (Japan Cedar). *Coniferæ.*

Increased by seeds, and by cuttings of growing wood planted in sandy soil, under glass.

Cubeba. See Piper.

Cucumber (*Cucumis sativus*). *Cucurbitaceæ.*

Seeds. If sown outdoors, the operation should be delayed until the weather is thoroughly settled.

Cucumber-tree. See Magnolia.

Cunninghamia (Broad-leaved China Fir). *Coniferæ.*

Increased by seeds, and cuttings of growing wood.

Cuphea. *Lythraceæ.*

Increased easily by seed ; but cuttings of the perennial sorts strike freely in spring, in brisk bottom heat.

Cupressus (Cypress). *Coniferæ.*

Seeds may be collected in early spring, and should be sown in April in a warm, friable soil. Cuttings of growing or mature wood, much as for Retinospora, which see.

Curculigo. *Amaryllidaceæ.*

Seeds ; also by suckers, which form at the base of the stem.

Curcuma (Turmeric). *Scitamineæ.*

Increased by root division.

Currant (*Ribes rubrum*, *R. nigrum* and *R. aureum*). *Saxifragaceæ.*

New varieties are grown from seeds, which may be sown in the fall or stratified until spring. Commercial varieties are nearly always multiplied by hard-wood cuttings (Fig. 65). The cuttings may be taken in spring and placed directly in the ground, but better results are obtained by taking them in the fall or late summer. Many nurserymen prefer to take them in August, strip off the leaves, and bury them in bunches with the butts up. They may remain in this condition or in a cellar all winter, or they may be planted in the fall. Currant cuttings strike readily, however, under any method. Some growers cut out the buds which stand below the surface of the ground, to prevent suckering, but this is not generally practiced ; the suckers are cut off when the cuttings are removed from the cutting-bed, either to be sold or to be transplanted into nursery rows. Strong plants, such as eastern markets demand, are usually obtained by allowing the cuttings to stand for two years before sale. Green layering is sometimes practiced with rare sorts, or single eyes may be used, as in grapes. Tip-layering, as in the black raspberry, may also be employed. (See page 36.) Weak or low sorts are sometimes grafted upon stronger ones, in order to give them a tree form, but such bushes are grown only as curiosities or as specimen plants.

Cussonia. *Araliaceæ.*

Increased by cuttings, which should be planted in sand, under glass. Give slight bottom heat.

Custard Apple. See Anona.

Cyananthus. *Campanulaceæ.*

Seeds. Strong roots may be carefully divided in spring, but this is not desirable. Usually by cuttings, which should be taken during spring or early summer, and struck in sandy peat, being kept moist.

Cyanophyllum. *Melastomaceæ.*

Increased by seed. By cuttings or eyes, which should be placed in sand, where a good bottom heat must be maintained, and they should be shaded from the sun.

Cyathea. See Ferns.

Cycas. *Cycadaceæ.*

Increased by seed, and oftener by suckers. Some, and perhaps all, of the cycads can be propagated by sections of the old stem or trunk. Cut the trunk into truncheons 2 or 3 inches thick, usually slanting; let the pieces dry a few days to guard against rotting, then plant in pots or sand. Roots will form between the scales, and new plants will push out. These should be removed and treated as independent plants. The severed crown of the trunk may also be potted, and it will grow.

Cyclamen (Sowbread). *Primulaceæ.*

Propagated by seed, sown when freshly gathered; the hardy kinds in pots placed in a cool frame. By divisions, and leaf cuttings taken off with a heel; but these methods are not very satisfactory.

Cypella. *Iridaceæ.*

Propagated by seed, sown as soon as ripe in a cool house, and by offsets.

Cyperus. *Cyperaceæ.*

Propagated either by seed, sown in gentle heat, or by divisions. *C. alternifolius*, the umbrella-plant, propagates readily from the crown or rosette of leaves. Cut off the crown, with an inch or two of stem remaining.

and set on sand or moss. Cut in the leaves. New plants will start from the axils. See Papyrus.

Cyphia. *Campanulaceæ.*

When the stems begin to push out from the root, cut off as many of the shoots as are required, and place them in small pots in an equal mixture of loam, peat, and sand in abundance. The young plants should be kept dry until callused, but not covered with glass. They may also be increased by cuttings, under a hand-glass in a cool house.

Cyphomandra (Tree Tomato of Jamaica). *Solanaceæ.*

Use seeds; or cuttings may be placed under glass, in bottom heat.

Cypress. See Cupressus.

Cypripedium (Lady's Slipper). *Orchidaceæ.*

By seeds sometimes; usually by divisions. (See under Orchids.)

Cyrilla. *Cyrillaceæ.*

Propagated by seeds and cuttings, like Andromeda, etc.

Cyrtanthus. *Amaryllidaceæ.*

Propagated by offsets.

Cytisus (Scotch Broom). *Leguminosæ.*

By seeds and layers. In spring, cuttings of young wood may be taken when about three inches long (with a heel preferred), placed under a bell-glass in heat, or in a close frame, where they will root readily. If gradually hardened, potted and grown on, small flowering specimens may be obtained the following spring. *C. purpurea* is usually grafted on the common laburnum. Species of Genista are propagated the same.

Dacrydium (Tear Tree). *Coniferæ.*

Increased by fresh seed and ripened cuttings.

Daffodil. See Narcissus.

Dahlia. *Compositæ.*

Commonly grown from tubers, which are dug in the fall and stored in the cellar, like potatoes. Each fork of the root may be broken apart and planted separately in the field; or the pieces may be started on early in pots or boxes.

Single varieties, and sometimes the doubles, are grown from seeds.

Dahlia tubers may be started into growth in heat late in winter, and the young sprouts may be removed and handled as ordinary cuttings as fast as they form, the same as sweet potatoes are handled. These cuttings should be removed close to the tuber or else at the first joint (preferably the former) and handled into small pots, where they will soon form tubers. These cutting-plants, if 6 to 10 inches high when set in the open, make excellent bloom that season, although generally giving dwarfer plants than those grown from tubers planted directly in the ground. Rare sorts may be increased during summer by cuttings from the growing tips. Cions made of the growing tips may be grafted into the roots by a cleft- or side-graft (see page 129). This method is oftenest employed for the purpose of preserving over winter rare sorts which it is feared may be lost. The grafts are kept growing slowly during winter, and cuttings may be taken from them. Sometimes cions are taken from forced plants in late winter or early spring and set in strong tubers for outdoor planting. Cuttings should always have a bud or buds at the base, and in propagation by division, there must be a piece of the crown attached to the root.

Daisy. See Bellis and Chrysanthemum.

Dalbergia. *Leguminosæ.*

Place cuttings of firm young shoots in sand under a glass, in spring. Give a little bottom heat.

Dandelion (*Taraxicum officinale*). *Compositæ.*

Seeds, in early spring, when grown for "greens."

Daphne. *Thymelæaceæ.*

Seeds. For layers, remove the soil in spring to a depth of 2 or 3 inches about the plant, and fill with fine compost to within two inches of the tops of the shoots. The next spring, carefully wash away the compost, and plant the small white buds in pots of fine soil. Place in a cool frame.

Cuttings should be made of matured shoots or side growths in autumn; insert thinly in well-drained pots of peaty soil, and cover with a bell-glass. If kept in a cool house in winter they will callus, and, early in spring, may be introduced to gentle heat, to encourage growth and the emission of roots. Pot the young plants singly, and grow on in a close but not high temperature, and afterwards harden and keep rather cool during the following autumn and winter, in order to thoroughly ripen the wood. Grafted specimens may be treated in a similar way. *D. odora* is propagated by ripened cuttings in a cool house, in sand. Sometimes the old wood can be used. The time is determined by the fitness of the wood.

Darlingtonia. *Sarraceniaceæ.*

Increased by seeds, and by dividing the plants. Seeds may be sown on the surface of well-prepared fibrous soil, and then covered with dead sphagnum moss, rubbed through a sieve. Give shade.

Dasylirion. *Liliaceæ.*

Increased by seeds, suckers and cuttings.

Date, Date Palm (*Phœnix dactylifera*). *Palmaceæ.*

The seeds from commercial dates grow readily, and without the intervention of stratification. Special varieties are propagated by a sort of cutting, made by removing and rooting the sprouts which appear about the base of the tree. These root readily if taken off green and liberally supplied with water. They often begin to bear in five or six years. The species grown for ornament are generally increased by suckers.

Datisca. *Datiscaceæ.*

May be increased by seeds, and by dividing well established plants.

Datura, including Brugmansia and Stramonium. *Solanaceæ.*

The annual species are propagated by seeds, which are started under cover in the north. The perennials are readily grown from cuttings in mild heat. Heeled shoots are usually preferred.

Davallia. *Filices.*

Propagated largely by division. See Ferns.

Day Lily. See Hemerocallis and Funkia.

Decumaria. *Saxifragaceæ.*

Seeds. Cuttings may be made in summer, and placed under a frame in a shady situation.

Delphinium (Larkspur). *Ranunculaceæ.*

Seeds may be sown outdoors in a warm border in spring, or in pans, to be placed either in frames or outside. The old plants of perennial sorts may be cut down after flowering, when young growths will spring from the base, and the whole may be lifted and carefully divided. Cuttings of the young shoots, taken in autumn or spring, will root freely if potted singly and placed in a coldframe. They will flower the following season at the same time as the divisions.

Dendrobium. *Orchidaceæ.*

Where a rapid increase of a new or special variety is required, the pseudo-bulbs that are more than one year old should be cut into lengths, and fastened on orchid rafts, with a layer of sphagnum beneath them. Suspend them in a hot, moist house, if possible, over a water-tank. The advantage of this method is that the young plants do not need shifting after they commence rooting on their own account. The section to which *D. aggregatum*, *D. Jenkinsii*, *D. densiflorum* and *D. thrysiflorum* belong are best propagated by division. (See under Orchids.)

Dentaria (Toothwort). *Cruciferæ.*

Propagated by seeds or divisions.

Deodar. See Cedrus.

Deutzia. *Saxifragaceæ.*

Commercially, the species are mostly propagated by green hardened cuttings in summer, under a frame. Hard-wooded cuttings may be taken in autumn, and be treated in about the same manner as currant cuttings (see pages 67, 68). The deutzias are also propagated by divisions and layers. Some of the dwarf sorts are sometimes forced, to make cuttings for winter use.

Dewberry (*Rubus Canadensis* and vars., *Rubus vitifolius* and *Rubus trivialis*). *Rosaceæ.*

Seeds are handled in the same manner as blackberry seeds. Increased by layers and, like the blackberry, by root cuttings. Layers are made by simply covering the decumbent canes at the joints. This is the usual method of multiplication. The tips, too, root freely, as in the blackcap raspberries, and it is from these that the commercial dewberry plants are mostly grown. See Blackberry.

Dianthera. As for Justicia.

Dianthus. See Carnation, Pink and Sweet William.

Dicentra, Dielytra (Bleeding Heart). *Fumariaceæ.*

The crowns may be divided in early spring, or cuttings may be made of the fleshy roots in short lengths, and placed in sand. The roots should be placed in a compost of sandy loam, in well-drained pots, as soon as the foliage dies off, and transferred to a coldframe. The native species propagate readily by the underground parts—*D. Cucullaria* by division of the bulbs, and *D. Canadensis* by the little tubers. All species grow from seeds which have been stratified.

Dichorisandra. *Commelinaceæ.*

Propagated by seeds, divisions and cuttings.

Dicksonia. *Filices.*

Division mostly. See Ferns.

Dictamnus (Dittany, or Fraxinella). *Rutaceæ.*

Seeds should be sown as soon as ripe. Division.

Dictyosperma. See Areca.

Didymocarpus. *Gesneraceæ.*

Cuttings, which are obtained from young shoots when commencing growth, and placed in sandy soil, in heat. Also by seeds.

Dielytra. See Dicentra.

Diervilla, Weigela. *Caprifoliaceæ.*

Suckers. Cuttings may be made in spring, summer or autumn. Hardened green cuttings, handled under a frame in summer, are extensively used by nurserymen. (See pages 67, 68.) They are sometimes grown from cuttings in winter from forced plants. Hard-wood cuttings, made in winter and planted in spring, like the grape, succeed well.

Dieffenbachia. As for Caladium.

Digitalis (Foxglove). *Scrophulariaceæ.*

Seeds, sown in spring, either indoors or in the open. The common foxglove (*D. purpurea*) often self-sows.

Dill (*Anethum graveolens*). *Umbelliferæ.*

Seeds, in early spring.

Dillenia. *Dilleniaceæ.*

Seeds, which, however, are grown with much difficulty. Cuttings of half-ripened wood may be placed in sand, under a frame, in bottom heat.

Dimorphanthus. See Aralia.

Dion, Platyzamia. *Cycadaceæ.*

Propagated by seed. See Cycas.

Dionæa. *Droseraceæ.*

Propagated sometimes by seed; usually by dividing the plants.

Dioscorea (Yam). *Dioscoreaceæ.*

The tubers may be divided in autumn or spring, when not growing. Start in heat. Seeds are sometimes used; so are the tubers which form in the axils by the leaves. Stove species can be propagated by cuttings of the half-ripened wood.

Diospyros (Date Plum, Persimmon). *Ebenaceæ.*

Seeds are used for the hardy species. Also by cuttings of half-ripened shoots. Those requiring stove heat strike best from ripened shoots, placed in sand in a brisk bottom heat during spring. See also Persimmon.

Dipladenia. *Apocynaceæ.*

In spring, when the plants commence new growth, cuttings from the young shoots are made. These, or single eyes, should be placed in a mixture of sand and peat in good bottom heat.

Diplothemium. *Palmaceæ.*

Propagation is effected by seeds.

Dirca. *Thymelæaceæ.*

Increased by seeds or layers.

Disa. *Orchidaceæ.*

D. grandiflora and others of similar habit are propagated by offsets. These are best taken off about December, and treated like the old plants. (See under Orchids.)

Disporum, including Prosartes. *Liliaceæ.*

Seeds may be used; or the plant may be divided in spring before active growth commences.

Dodecatheon (American Cowslip). *Primulaceæ.*

Seeds. The crowns may be divided either in spring or autumn. Cuttings of the whole root can be effectively used, the root being torn off the crown, planted upright, and covered with the sandy soil commonly used in this form of propagation.

Dog's Bane. See Apocynum.

Dog's-tooth Violet. See Erythronium.

Dogwood. See Cornus.

Dolichos. *Leguminosæ.*

By seeds. Sometimes cuttage or layerage is resorted to.

Doronicum (Leopard's Bane). *Compositæ.*

Propagated by seeds and divisions.

Dorstenia. *Urticaceæ.*

Seeds may be sown in a hotbed in early spring. Before active growth commences, the plants may be divided.

Doryanthes. *Amaryllidaceæ.*

Propagated by suckers placed in small pots.

Downingia, Clintonia. *Lobeliaceæ.*

Seeds should be sown in mild heat in spring.

Draba (Whitlow Grass). *Cruciferæ.*

The annuals or biennials are propagated by seeds sown in spring in the open border. The perennials may be propagated by dividing the crowns.

Dracæna. See Cordyline.

Dracocephalum (Dragon's Head). *Labiatæ.*

The annuals are grown from seeds, sown in the open in spring. Perennials are increased by dividing the roots, or by cuttings of the young shoots in spring.

Dracontium. See Amorphophallus.

Dragon-tree. See Cordyline

Drimys, Wintera. *Magnoliaceæ.*

Cuttings made of half-ripened shoots should be inserted in a frame. Seeds, when obtainable.

Drosera (Sundew). *Droseraceæ.*

Seeds, sown as soon as possible after gathering. *D. binata* is increased by cutting roots from strong plants into pieces of one-half or one inch in length, and placing them on the surface of shallow earthenware pans, in sandy peat soil, and covering about one-half inch deep with the same material. They are then placed under a bell-glass, and transferred to a damp, warm propagating house. This will suggest treatment for other species.

Drosophyllum. *Droseraceæ.*

Propagated by seed.

Dutchman's Pipe. See Aristolochia.

Eccremocarpus (Calampelis). *Bignoniaceæ.*

Seeds, sown in spring, in a gentle heat. Cuttings may be used of green or ripe wood.

Echeveria. See Cotyledon.

Echinacea. *Compositæ.*

Readily propagated by seeds and division.

Echinocactus. See Cactus.

Echinops (Globe Thistle). *Compositæ.*

Sow the seeds in spring for the propagation of the biennials, and divide the perennials early. Also by root cuttings.

Edelweiss. See Leontopodium.

Egg-Plant (*Solanum Melongena*). *Solanaceæ.*

Seeds in heat, in late winter or spring. Cuttings rarely.

Eglantine. See Rosa.

Eichhornia (*Pontederia azurea* of florists). *Pontederiaceæ.*

Propagation is effected by division in spring; seeds.

Elæagnus (Oleaster, Wild Olive, Goumi). *Elæagnaceæ.*

Increased by seeds, layers or cuttings. Hard-wood cuttings of *E. hortensis* strike readily. The named varieties are often grafted on the most vigorous varieties obtainable. Imported seeds of some species are apt to be empty. *E. longipes* can readily be propagated by cuttings of the half-ripened wood in June and July, under glass.

Elder. See Sambucus.

Elecampane (*Inula Helenium*). *Compositæ.*

Propagated by seeds in open air in early spring; but generally by division of the stools.

Elm. See Ulmus.

Empetrum (Crowberry, or Crakeberry). *Empetraceæ.*

Seeds. In summer, cuttings may be made, and should be placed in sandy soil under glass.

Encephalartos. *Cycadaceæ.*

Increased by seeds. See Cycas.

Endive (*Cichorium Endivia*). *Compositæ.*

Seeds, either in the open where the plants are to stand, or under glass.

Eomecon. *Papaveraceæ.*

Seeds; also by division.

Epacris. *Epacridaceæ.*

Grown from tip cuttings in a frame in winter, with bottom heat. The cuttings root very slowly.

Ephedra. *Gnetaceæ.*

Layers may be made from young shoots or branches.

Epidendrum. *Orchidaceæ.*

The tall-stemmed section of this genus is increased by cuttings, the section with short, thick pseudo-bulbs by division. The former also occasionally produces viviparous flower-scapes, thus affording a ready means of increase. (See under Orchids.)

Epigæa (Trailing Arbutus). *Ericaceæ.*

Increased with great difficulty by careful divisions of established plants, and by layers. Seeds, when obtainable, can be used, but are slow to develop. Cuttings are most successful. Use last year's wood in house in winter, putting them in sand. Pot them up as soon as established, and keep them in pots until they are set into permanent quarters.

Epimedium (Barrenwort). *Berberidaceæ.*

Sometimes increased by seeds. During July or August, divisions of the roots can be made.

Epiphyllum. *Cactaceæ.*

Readily grown from cuttings. Pieces of the branches 4 to 6 inches long are placed in sandy soil in gentle heat, and kept moderately dry. Epiphyllums are often grafted on strong stocks of pereskia (*Pereskia aculeata* is commonly used, but *P. Bleo* is equally as good), for the purpose of getting high or rafter plants. A young shoot is cleft- or side-grafted into any portion of the pereskia which has become hard, and the cion is held in place by a cactus spine passed through it. Several cions may be inserted along the sides of the stock. See Cactus.

Eranthemum. *Acanthaceæ.*

Seeds. Cuttings root readily in spring in peaty soil, in a close frame where there is a bottom heat of about 70°.

Eranthis (Winter Aconite). *Ranunculaceæ.*

Increased by seeds and divisions.

Eremurus. *Liliaceæ.*

Increased by seeds and divisions.

Erica (Heath). *Ericaceæ.*

Will grow from seeds, but these are used generally to secure new varieties. If seeds are employed, sow on peat or live sphagnum, and exercise great care not to let them dry out. Commonly propagated by very short cuttings, taken from the tips, or made of the lower young growth. Carefully remove the leaves from the lower parts of the cutting, which should be about one inch long, and then insert rather closely in pots, which should be filled two-thirds with crocks, the remainder being fine sandy peat with a layer of clean, compact sand on the surface. Cover with glass. Water well, and place in a temperature of about 60°.

Erinus. *Scrophulariaceæ.*

Seeds and divisions. After becoming established, they propagate themselves by seeds.

Eriobotrya. See Photinia.

Eriodendron. *Malvaceæ.*

Raised from seeds sown in sandy soil, in heat.

Eriostemon. *Rutaceæ.*

Cuttings, in sandy peat in spring, under glass, and with gentle heat. Nurserymen propagate by grafting on small stocks of correa.

Erodium (Heron's Bill). *Geraniaceæ.*

By seeds or division.

Eryngium (Eryngo). *Umbelliferæ.*

Seeds or carefully made divisions may be used for increasing the species.

Erysimum (Hedge Mustard). *Cruciferæ.*

Increased by seeds; the perennials by seeds and divisions.

Erythræa (Centaury). *Gentianaceæ.*

Propagated by seeds or division.

Erythrina (Coral-tree). *Leguminosæ.*

Seeds. Cuttings of young shoots can be taken in spring or early summer with a heel, and placed in sandy soil, on a slight bottom heat.

Erythronium (Dog's-tooth Violet). *Liliaceæ.*

Seeds. Offsets or bulbels are usually employed, taken as soon as the leaves dry away after flowering, inserting the bulbels about three inches deep.

Erythroxylon. *Linaceæ.*

Place cuttings of half-ripened shoots in sand under a glass, in heat. Seeds, if obtainable.

Eschscholtzia (California Poppy). *Papaveraceæ.*

Seeds may be sown in spring or autumn where the plants are to flower.

Eucalyptus (Gum-tree). *Myrtaceæ.*

Increased by seeds, which, for culture under glass, should be sown thinly in pans or pots of light, sandy soil, and placed in frames. Also by cuttings.

Eucharis. *Amaryllidaceæ.*

Seeds may be sown as soon as ripe in a warm house. Offsets or bulbels should be removed and potted off singly.

Eucomis. *Liliaceæ.*

Increased by seeds, sown as soon as ripe, or by bulbels.

Eugenia. See Myrtus.

Eulalia. See Miscanthus.

Euonymus (Burning-bush, Strawberry-tree, Wahoo). *Celastraceæ.*

Grown from seeds, cuttings and layers. Cuttings usually make better plants than layers. The deciduous species are usually grown from hard-wood cuttings, but the evergreen kinds are started under glass, from cuttings of the growing or ripened wood. The small and weak kinds are grafted on the stronger ones. The evergreen species will grow upon the deciduous kinds.

Eupatorium (Boneset). *Compositæ.*

Cuttings of the growing wood, under glass in early spring, is the common method of propagation of the conservatory species. Seeds can also be used for some species

Euphorbia, including Poinsettia (Spurge). *Euphorbiaceæ.*

By seeds, especially the annual species. The perennial shrubby sorts are increased by cuttings in a strong heat. Some species are propagated by divisions. See Poinsettia.

Eurycles. *Amaryllidaceæ.*

Offsets or bulbels, in spring.

Euterpe. *Palmaceæ.*

Seeds in heat.

Eutoca. See Phacelia.

Evening Primrose. See Œnothera.

Exochorda. *Rosaceæ.*

Grown from seeds, layers, cuttings and suckers. Layering in June is a common practice. Various kinds of cuttings are employed, but the best results follow short, soft cuttings, taken from forced plants and set deep in shallow flats of sand. They require a very strong bottom heat, a close frame, and the water should be applied in a spray upon the foliage. Cuttings are sometimes grafted upon pieces of roots. It has been regarded as a difficult plant to propagate, but seeds are now easily procured from cultivated plants, and they grow readily

Fagus (Beech). *Cupuliferæ.*

Commonly grown from the nuts, which should be stratified and sown very early in spring. They may be sown immediately after they are gathered, if they can be protected from vermin. The named varieties are grafted upon the European or American species. (See Fig. 115 for a good method.) The purple-leaved beech reproduces itself very closely by seeds, although different shades of purple will appear amongst the seedlings.

Farfugium. See Senecio.

Felicia. *Compositæ.*

Propagated by seeds, or by cuttings inserted in sandy soil, under glass.

Fennel (*Fœniculum*, various species). *Umbelliferæ.*

Seeds, usually in spring.

Fennel Flower. See Nigella.

Fenugreek (*Trigonella Fœnum-Græcum*). *Leguminosæ.*

Propagated by seeds.

Ferns. *Filices.*

Where division is possible, it is the easiest and most economical method of propagation, and should be practiced just before the plant starts into growth. The spores can be sown in February and March, or earlier, under glass, in a warm propagating pit. Partly fill a suitable sized pot or pan with coarse peat, giving plenty of drainage; make the surface level, and on this place three-quarter inch cubes of well-seasoned peat which is rather dry, watering the whole and scattering on the spores evenly. Cover with a pane of glass, and place in a partial shade. While the process which corresponds to germination is going on, great care must be given to the water supply. This is sometimes done by placing the pots or pans in a saucer, from which they can suck the water up. Overhead watering may be used, and often is, but it must be done with great care. Be certain that the spores are fully ripe when gathered.

The young plants should be pricked out when the true leaf appears, and they are large enough to handle. The same careful treatment should be continued until they are established in pots.

There is not much difficulty in getting the young plants, if fresh spores are obtainable, but there is a good deal of trouble in handling the plantlets, and establishing them in their growing quarters.

Most ferns are readily propagated by means of spores, as directed above and on page 24. Some species rarely produce spores in cultivation, however, and in other cases, as in some tree ferns, it is almost impossible to rear the young plants after the spores have germinated. In all such cases, recourse must be had to separation, division or layerage. There are some species, as *Asplenium bulbiferum*, *Cystopteris bulbifera* and others, which bear small bulblets or detachable buds on their fronds. These buds often vegetate while still attached to the frond. They may be removed either before or after showing signs of vegetation, and set in pots in a close propagating frame, or under a bell-glass. Ferns which make broad crowns may be divided, and this is the common mode with many species. Some species produce creeping rootstocks, which emit roots if pegged down into a pot of soil or on a block of peat. Several plants can often be produced from such a layer. All these operations are best performed in late winter, before the new growth begins. The tree ferns are rarely propagated to any extent in cultivation, but young plants are imported from their native countries.

Fern, Sweet. See Comptonia.

Ferraria, Tigridia. *Iridaceæ.*

Propagated by means of seeds and bulbels.

Feverfew. See Chrysanthemum.

Ficus. *Urticaceæ.*

The greenhouse species are propagated by layers and cuttings. The cuttings are handled in a close frame, and a leaf or two is usually left on them. For *Ficus Carica*, see Fig. Propagation by seeds is sometimes used in the edible figs, but is not easy with the ornamental sorts. *F. elastica*, *F. Indica*, etc., are increased by cuttings (commonly single-eye), planted in sand or sandy soil or sphagnum, and placed in good bottom heat, in a frame under glass. The large cuttings should be staked, and care must be taken to remove the milky juice before planting. Any winter month is good, before growth begins. Last season's wood should be used. A common method of multiplying *F. elastica* (Rubber-plant) is by means of Chinese or air layers (see page 41). If the house can be kept moist, simply a ball of sphagnum bound on the stem is sufficient, without the use of a split pot or a paper cone (as shown in Figs. 33 and 35). Plants of considerable size, fit for nursery trade, can be obtained quicker by this Chinese layering (if one has good stock plants) than by cuttings.

Fig (*Ficus Carica*). *Urticaceæ.*

Figs grow readily from the plump seeds in the commercial fruit. Wash out the seeds, and those that sink may be sown in a frame. The young plants will appear in three or four weeks. In from three to five years the plants will begin to bear. New varieties are obtained in this way.

Varieties of the fig are multiplied with ease by layers, suckers and cuttings. Make cuttings of mature wood in autumn, cutting just below a bud. Scarce varieties may be multiplied by single-eye cuttings. Fig cuttings are handled in the same way as grape cuttings. Some prefer, however, to place the cuttings where the tree is to stand. A well-grown plant will bear at two or three years of age.

The fig is readily budded and grafted, but these methods are seldom employed, because the plant is so easily multiplied by cuttings. Shield, ring or tubular buddings are employed. Various methods of grafting are adapted to it, and cleft-grafting is usually employed on old plants.

Filbert. See Corylus.

Fir. See Abies, Picea and Pinus.

Fire-pink. See Silene.

Fittonia. *Acanthaceæ.*

Increased by division, and by cuttings of half-ripened shoots, planted in sandy loam, in bottom heat.

Fitzroya. *Coniferæ.*

Seeds. Increased also by cuttings of half-ripened shoots.

Flax. See Linum.

Flower-de-Luce (*Fleur-de-Lis*). See Iris.

Fontanesia. *Oleaceæ.*

Layers are used; also cuttings, planted under a handglass in autumn. Or it may be grafted on the privet.

Forget-me-not. See Myosotis.

Forsythia (Golden Bell). *Oleaceæ.*

Propagated extensively by green cuttings in summer, in a frame; also grown from ripe cuttings taken in fall and winter, and planted in the open air in early spring.

Fothergilla. *Hamamelideæ.*

Propagated by seeds, sown in spring in a peaty soil; by layers.

Four-O'clock. See Mirabilis.

Foxglove. See Digitalis.

Fragaria. See Strawberry.

Franciscea. See Brunfelsia.

Francoa. *Saxifragaceæ.*

Seeds, sown in early spring in a cool frame. Also by division.

Frangula. See Rhamnus.

Fraxinella. See Dictamnus.

Fraxinus (Ash). *Oleaceæ.*

Propagated chiefly by seeds, which should be stratified until fall or the spring following the gathering. The seeds do not germinate the year in which they mature. The named sorts are budded upon seedling stocks if the sorts are upright growers, or top-grafted if they are weepers. Both the European and American species are used for stocks.

Freesia. *Iridaceæ.*

Increased readily by seeds, sown as soon as ripe in pots of light, sandy soil, and placed in a sunny position, in a cool frame.. Commonly by bulbels.

Freycinetia. *Pandanaceæ.*

Increased by offsets. Seeds, when obtainable.

Fringe-tree. See Chionanthus.

Fritillaria. *Liliaceæ.*

Seeds, sown as soon as ripe where the plants are to stand the first year. Bulbels and division.

Fuchsia (Ladies' Ear Drop). *Onagraceæ.*

Fuchsias grow readily from seeds, which should be sown as soon as ripe, and blooming plants ought to be obtained in eight or ten months. Cuttings of the young growth strike quickly and easily. Blooming plants of most sorts can be obtained in four or five months. Plants for winter bloom are usually started in late spring.

Funkia (Plantain Lily, White Day Lily). *Liliaceæ.*

Propagation is effected by dividing the stools during the early autumn, or when they begin to start in spring. Only strong, healthy clumps should be divided, and each portion should contain several crowns.

Furze. See Ulex.

Gaillardia. *Compositæ.*

The annual sorts are propagated by seeds started under glass; the perennial kinds by seeds, cuttings or division. Sometimes root cuttings are used.

Galanthus (Snowdrop). *Amaryllidaceæ.*

Commonly by bulbels. Rarely by seeds.

Galax. *Diapensiaceæ.*

Propagated by divisions of strong clumps in autumn.

Galega (Goat's Rue). *Leguminosæ.*

Seeds, in spring; also by division.

Galtonia (*Hyacinthus candicans* of gardeners). *Liliaceæ.*

Increased by bulbels or seeds.

Garcinia, Cambogia, Mangostana. *Guttiferæ.*

Seeds. Cuttings of ripened shoots should be inserted in sand under a glass, in strong bottom heat.

Gardenia. *Rubiaceæ.*

Strong, healthy cuttings may be taken with a heel, early in the year being the best time, but any season will do when suitable cuttings can be secured. They should be placed in bottom heat of about 75°, in a frame.

Garlic (*Allium sativum*). *Liliaceæ.*

By "cloves" or divisions of the bulb. In the north these are planted in the spring, but in warm climates they may be planted in the fall.

Garrya, including Fadyenia. *Cornaceæ.*

Propagated by seeds, or by cuttings of half-ripened wood in sandy loam in August, and shaded until rooted. Also by budding on *Aucuba Japonica* at the crown. Plant sufficiently deep to cover the bud or graft.

Gaultheria (Boxberry, Wintergreen). *Ericaceæ.*

Increased by seeds, divisions, layers and cuttings under glass.

Gaylussacia. See Vaccinium and Whortleberry.

Gazania. *Compositæ.*

Increased by seeds, and by divisions. Make cuttings in July or August, from the side shoots near the base of the plant; these should be placed in a sandy soil, in a frame.

Gelsemium. *Loganiaceæ.*

Propagated by cuttings under glass.

Genista. See Cytisus.

Gentiana (Gentian). *Gentianaceæ.*

Seeds and division. The seeds germinate slowly, and often with difficulty. They often lie dormant a year or more. They should be sown in well-sifted light loam, in pans or flats, and kept cool and shaded. Division must be carefully done, or the plants will suffer.

Geonoma. *Palmaceæ.*

Increased by seeds and suckers.

Geranium. *Geraniaceæ.*

Mostly by seeds and divisions. For the conservatory plants known as geraniums, see Pelargonium

Gerardia. *Scrophulariaceæ.*

Propagated, but often with difficulty, by seeds, sown in the open air or in a frame or cool house. Many of the species are partially parasitic on roots.

German Ivy. See Senecio.

Gesnera. *Gesneraceæ.*

Seeds, and cuttings of the shoots and leaves. Handled in essentially the same manner as Sinningia, which see.

Gethyllis. *Amaryllidaceæ.*

They may be increased by bulbels or seeds.

Gherkin (*Cucumis Anguria*). *Cucurbitaceæ.*

Propagation is effected by seeds. See Cucumber.

Gilia, including Fenzlia. *Polemoniaceæ.*

Seeds should be sown in spring in the open ground or frame, in a rather light soil.

Gillenia. *Rosaceæ.*

Increased readily by dividing the roots in spring; also by seeds.

Gilliflower. See Matthiola.

Ginger. See Zingiber.

Ginkgo, Salisburia (Maidenhair-tree). *Coniferæ.*

Seeds, which are mostly imported, and which should be stratified. Seeds are now produced in some quantity in this country. Also by layers, and by cuttings of either green or ripe wood. The cuttings are handled under glass. Named varieties are grafted upon common stocks.

Ginseng (*Aralia*, or *Panax*). *Araliaceæ.*

Cuttings of stems and roots. Stems of old plants may be cut into pieces an inch or two long and inserted in sand in heat. Or young plants can be obtained by cutting down the tops of strong plants and then separating the suckers which arise.

Gladiolus. *Iridaceæ.*

Seeds, which are commonly sown in pans in spring, in the house; or they may be sown in the border. Seedlings flower in two or three years. They give new varieties. The common method of propagation is by means of cormels (see page 31, and Fig. 25). These are removed from the parent corm and planted in the open, where some of them will flower the same season, although most of them will require a season's independent growth before they flower. If cormels are desired in abundance, the large corms should not be allowed to flower. Some varieties do not produce cormels readily, and these may be made to bear them by cutting or ringing (page 31). One or more new corms are formed above the old one each year (Fig. 25).

Gleditschia (Honey Locust). *Leguminosæ.*

Seeds should be sown in spring about one inch deep. They should be soaked in hot water before being sown. Varieties propagated by grafts upon seedling stock.

Gleichenia. See Ferns.

Globe Flower. See Trollius.

Gloriosa. *Liliaceæ.*

Seeds should be inserted singly in small pots, in a

light, sandy soil, and plunged in bottom heat. Bulbels, which should be carefully removed from the old bulbs when starting them in spring, as the roots are very brittle.

Gloxinia. See Sinningia.

Glycosmis. *Rutaceæ.*

Seeds. Increased by cuttings, which are commonly inserted in sand under glass, often in heat.

Glycyrrhiza (Liquorice). *Leguminosæ.*

Propagated by division and by seeds.

Godetia. See Œnothera.

Golden Rod. See Solidago.

Gomphocarpus. *Asclepiadaceæ.*

Seeds should be sown under glass in spring; or cuttings may be made of small side shoots when the plant is commencing new growth, and placed in sand under glass.

Gomphrena. See Celosia.

Gonolobus. *Asclepiadaceæ.*

Seeds, divisions, and cuttings under glass.

Goober. See Arachis.

Gooseberry (*Ribes Grossularia* and *R. oxyacanthoides*). *Saxifragaceæ.*

Seeds, for the raising of new varieties, should be sown as soon as well cured, in loamy or sandy soil, or they may be stratified and sown together with the sand in the spring. Cuttings 6 to 8 inches long of the mature wood, inserted two-thirds their length, usually grow readily, especially if taken in August or September and stored during winter, in the same way as currant cuttings (Fig. 65). Single eye cuttings may be used for rare kinds. Stronger plants are usually obtained by layers, and the English varieties are nearly always layered in this country (although frequently grown from cuttings in England). Mound-layering is usually employed (the mounding being done in June, or when the new growth has reached several inches), the English varieties being allowed to remain on the stools two years, but the American varieties only one (Fig. 32). Much depends upon the variety. The Downing, for example, usually makes a merchantable plant in one year after transplanting from the stools, but Smith Improved may require a year more. Layered plants are usually set in nursery rows for a year after removal from the stools. Green-layering during summer is sometimes practiced for new or rare varieties. Strong plants may also be procured by tip-layering, as in the black raspberry (see page 36). If it is desired to train the weaker gooseberries in tree form, they may be grafted upon the stronger-growing varieties.

Gordonia. *Ternstræmiaceæ.*

Propagated by seeds or layers.

Gorse. See Ulex.

Goumi. See Elæagnus.

Gourds (*Cucurbita Pepo*, *Lagenaria*, etc.). *Cucurbitaceæ.*

Seeds, after the weather is settled and ground is warm.

Granadilla (*Passiflora edulis*, etc.) *Passifloraceæ.*

Propagated by seeds, or, less easily, by cuttings.

Grape (*Vitis*, several species). *Vitaceæ.*

Grape seedlings are very easily grown. If the ground is fit and there is no danger from vermin, the seeds may be sown in the fall, but they are usually stratified and sown in spring. They come readily if sown outdoors, but some prefer to force them under glass with a mild bottom heat. Seedlings do not "come true," and they are therefore grown only for the purpose of obtaining new sorts.

The grape is very readily multiplied by layers, either of the ripe or green wood. The ripe wood or canes may be layered either in fall or spring, but spring is usually chosen. The cane is simply covered up 2 or 3 inches deep, and nearly every bud will produce a plant. In August or September the layer should be lifted and cut up into plants. Better plants are obtained if only the strongest canes are used and only a part of the buds on each are allowed to grow. The cane is usually cut back to four or five buds, or if very strong plants are desired only one bud is left on each layer. Canes of the previous year, those recently matured, are preferred, although wood two or three years old may be used, but in this case it is usually necessary to cut or otherwise wound the joint in order to induce the formation of roots. Vines or stools grown for the production of layers should be cut back severely in fall or winter, to induce a vigorous growth of canes the

Grape, continued.

following season. These canes are then layered the succeeding fall or spring. Only a part of the canes are layered from any stool, a part being allowed to grow for cutting back the next fall in order to get another crop of canes. In some varieties which do not strike readily from cuttings, layering is considerably practiced by nurserymen. The Delaware is often grown in this way. Extra strong layers can be secured by layering in pots. A large pot, filled with rich soil, is plunged beneath the layer. In this manner a layer may be rooted and separated even while carrying fruit. Layering in pots is employed only in special cases. In vineyards, layering is often employed for the purpose of filling vacancies. A strong cane is left, without pruning, on a neighboring vine in the same row, and in the spring the end of it is laid down in the vacant place. The vine is covered about a half foot deep, and the free end of it is turned up perpendicularly out of the soil and tied to a stake. By fall or the following spring the layer should be sufficiently rooted to allow the parent cane to be cut away.

Green-layering is sometimes practiced upon new and scarce varieties, but strong plants are not obtained unless they are well handled by forceful culture after they are separated. The growing cane is layered in midsummer, usually by serpentine layering.

Cuttings are usually employed by nurserymen to propagate the grape. These are made in many fashions. In all ordinary cases hard-wood cuttings are made from the ripened canes in autumn or winter when the vines are pruned. It is advisable to take the cuttings before the canes have been exposed to great cold. Select only those canes which are well matured, solid and rather short-jointed. In common practice, the cuttings are cut into two-bud lengths, the lower cut being made close to the bud. The cuttings will range from 6 to 10 inches in length. Some prefer three-bud cuttings (Fig. 59), but unless the cane is very short-jointed, such cuttings are too long to be planted and handled economically. Three-bud cuttings usually give stronger plants the first season, because roots start from both joints as a rule. Very strong plants are obtained from mallet cuttings (Fig. 61), but as only one such cutting can be made from a cane, unless the cane bears very strong branches, they are not much used. Various methods of peeling, slitting and slicing cuttings are recommended, in order to extend the

Grape, continued.

callusing process, but they are not used in common or commercial practice. The cuttings are tied in bundles of 50 or 10 , and stored in sand, moss, or sawdust in a cellar, until spring, when they are planted in rows in the open. Some varieties, of which the Delaware is an example, do not strike readily from cuttings. Some growers start common cuttings of these under glass in spring. Others bury the bundles of cuttings in a warm exposure in the fall, with the butt ends up and about level with the surface of the ground. This affords bottom heat to the butts and induces callusing. (See page 57.) At the approach of cold weather the cuttings are removed to a cellar, or are heavily mulched and allowed to remain where buried. Storing is safer. Some growers obtain the same results by burying upside down in a cellar. These slow-rooting sorts often start well if they are simply kept in a warm cellar—but where the buds will not swell—all winter, as the callusing is then hastened. At the end of the first season the plants may be transplanted. The plants are often sold at this age, but buyers usually prefer two-year-old plants.

Single bud or "eye" cuttings are largely used for the newer and rarer varieties. These are cut from the canes in the fall, the same as long cuttings, and are stored in boxes of sand or moss. A month before the weather becomes settled, these boxes may be taken into a house or greenhouse, or put in a mild hotbed, to induce the formation of the callus. They may then be planted outdoors, and a fair proportion of most varieties may be expected to grow. The best and commonest way of handling eyes, however, is to start them under glass. They are planted horizontally, or nearly so, and about an inch deep in sand or sandy earth, in a cool greenhouse in late winter—in February in the northern states—and in about six weeks the plants will be large enough to pot off or to transplant into coldframes or a cool house. If only a few plants are to be grown, they may be started in pots. When the weather is thoroughly settled, they are transferred to nursery rows, and by fall they will make strong plants. There are various ways recommended for the cutting of these eyes—as cutting the ends obliquely up or down, shaving off the bark below the bud, and so on—but the advantages of these fashions are imaginary. A good eye-cutting is shown in Fig. 66. The foreign grapes are propagated by eyes in the north.

Soft cuttings are sometimes used to multiply new kinds.

Grape, continued.

These may be taken in summer from the growing canes, but the plants are usually forced during winter for the purpose of giving the extra wood. Cuttings are taken off as fast as buds form during the winter, and they are forced in close frames with a good bottom heat. The cuttings may comprise two buds, with the leaf at the upper one allowed to remain, or they may bear but a single eye, in which case the leaf, or the most of it, is left on. This rapid multiplication from small, soft wood usually gives poor plants; but strong plants may be obtained by allowing the wood to become well hardened before it is used. Soft cuttings will root in two or three weeks under good treatment.

In order to secure extra strong plants from single buds, the eyes may be saddle-grafted or whip-grafted upon a root 2 or 3 inches long. The root grafts are then treated in the same way as eye cuttings, only that they are usually grown in pots from the start.

The vine may be grafted with ease by any method. Cleft-grafting is commonly employed upon old plants. The cions are inserted on the crown of the plant, three or four inches below the surface of the ground. The cleft is bound with string, and then covered with earth, no wax being necessary. The best time to perform the operation is very early in spring, before the sap starts. Vines may be grafted late in spring also, after danger of bleeding is past, if the cions are kept perfectly dormant. Vines are sometimes grafted in the fall, but this practice cannot be recommended in the north. Young plants are usually whip-grafted at the crown, either indoors or outdoors. Grafting the vine is mostly confined to Europe, California, and other countries where the European grape (*Vitis vinifera*) is grown, as that species must be grafted upon some other stock in order to resist the phylloxera. The common wild frost-grape (*Vitis riparia*) is the most popular stock. The union in these cases must be two or three inches above the ground, to prevent the cion from taking root. The union is wound with waxed muslin, and the earth is heaped about it until it has healed. Grapes can be grafted by the cleft-graft below ground as readily as pears or apples can be worked. For pictures of various methods of grafting the grape, see Figs. 107, 113, 114, 116, 124, 125. The last (125) is the best type for general use on old vines.

The vine is frequently inarched, and early in spring it can be budded by ordinary methods.

Grape, continued.

Seed-grafting is a curious practice, which may be applied to the grape (see page 131).

There is so much misapprehension respecting the methods and results of the grafting of grapes, that the following directions by the veteran viticulturist, George Husmann, now of Napa, California (as given in American Agriculturist, 1896), are here transcribed in full:

"A good, thin-bladed, sharp knife to cut the cions, a sharp saw to cut off large stocks—the smaller ones can be cut with good pruning shears—a chisel for grafting having a blade 2½ or 3 inches broad in the middle and a wedge on each side [a knife with but a single wedge, as in Figs. 120 and 121, will answer the same purpose], a wooden mallet, and a few strings of raffia, or other bandage, in case a stock should need tying, which is seldom the case—are the implements necessary for grafting. The cions should be of selected wood, the size of a lead-pencil, or somewhat larger, cut some time in winter, tied in bundles, and buried their entire length on the shady side of a building, or under a tree, to keep them dormant. Short-jointed, firm wood is to be preferred. All can be carried in a basket, if one intends to perform the operation alone. If several are to work together, of course the tools must be divided accordingly. In California we work generally in gangs of three, the first man clearing away the ground from the stock until he comes to a smooth place for inserting the cion, whether this be at the surface or slightly below. The former is preferable if resistant vines are to be grafted with non-resistant cions. He then cuts off the stock horizontally about an inch and a half above a knot or joint. The next man cuts the cions to a smooth, long, sloping wedge just below a bud [as shown in Fig. 119], then splits the stock, either with pruning shears or chisel, according to its size. If the stock is not more than an inch in diameter, the shears are best, as only one cion is to be inserted. Keep the blade of the shears on the side where the cion is to join the stock, so as to prevent bruising, and make a long, smooth, sloping cut, a little transversely if possible, as the junction will thereby become all the more perfect. Then push the wedge of the cion firmly down into the cleft, taking care that the inner bark or fiber of stock and cion are well joined, as on this principally depends the success of the operation. To open the cleft, the wedges on the chisel are used if necessary. An expert will depend very little

Grape, continued.

on these, unless the stocks are very heavy, but will open the cleft with knife or shears, and then push down the cion to its proper place. The inner side of the cion, opposite the bud, should be somewhat thinner, so that the stock will close firmly on it; the cion should also be inserted far enough so that the bud is just above the horizontal cut on the stock. The third man follows, presses a little moist earth on the surface of the stock, and then hills up around the junction to the uppermost buds of the cion with well-pulverized soil, taking care not to move the cion, and the operation is finished. It becomes necessary sometimes to tie the stock, when it is not large enough or from some defect in grafting it does not firmly hold the cion. In such a case, pass a string of raffia or some other flat bandage firmly around the stock and tie it, but in no case use grafting wax or clay, as the strong flow of sap from all the pores is apt to drown and sour the cion, while without obstructing it, it will flow around the stock, serving to keep the junction moist and facilitate the union. As the whole operation is covered with earth, there is no danger of drying up, as is sometimes the case when fruit trees are top-grafted.

"A very important consideration, to insure success, is to equalize the stock and cion. If, therefore, large stocks are to be grafted, we must have strong, well-developed wood for the cions, and have buds enough to take up the full flow of sap, while small stocks, if used at all, should be grafted with small cions of only two or three buds. When the stocks are strong, I take two cions and insert one on each side of the stock, of full length, say from 14 to 16 inches, and with six to eight buds each. This has many advantages. The principal one is that they will elaborate and work up the entire flow of sap. Another is, that if the cions have well-developed fruit buds, they will produce quite a number of clusters from the upper buds, and thus show the character of the fruit the first year. I have picked a thousand pounds of grapes from an acre thus grafted, the first summer, and a full crop of five or six tons per acre the following season. Another advantage is that it establishes the crown of the graft at the right distance from the ground, as the three upper buds will produce the canes for the next season's bearing. If both cions grow, cut off the weakest above the junction the next spring, leaving only the strongest. I generally find that the whole surface of the stock is covered by the

Grape, concluded.

new growth, and that the junction between stock and cion is perfect. Another advantage is—especial'y in California, where we plow and cultivate close to the vines, and where some of the workmen are careless—they are more apt to run over and disturb the small grafts than the large ones, which are protected by hills of earth above the surface; nor are the young shoots disturbed and broken so easily by careless hands or high winds. A stake should be driven close to the graft immediately after grafting is finished, and the young shoots, when they appear, tied to it for support, as they generally start vigorously and are easily broken off, or blown off by high winds. Do not be discouraged if some time elapses before they start. I have often had them remain dormant until July or August, and then make a rapid growth. If suckers from the stock appear—as is generally the case—they should be removed at once, taking care to cut them close to the stock, so as to have no stumps or dormant buds. Tying and suckering should be repeated every week or ten days at least. As long as the cion remains fresh and green it may begin growing at any time. Of course, care must be taken not to disturb the cion. If everything does well, there will be three or four canes from the upper buds, which may be pruned just as any other bearing vines."

Grape Hyacinth. See Muscari.

Grevillea. *Proteaceæ.*

Propagated by seeds, sown under glass in late winter; also by cuttings of half-ripened wood.

Grewia. *Tiliaceæ.*

Seeds. Cuttings may be struck in sand under glass, with heat.

Grindelia. *Compositæ.*

Seeds, sown in the border or under a frame. Divisions. Cuttings.

Ground-Cherry. See Physalis.

Ground-Nut. See Apios; also Peanut (under Arachis).

Groundsel. See Senecio.

Guaiacum. *Zygophyllaceæ.*

Ripened cuttings in spring, under a hand-glass, in heat.

Guava (*Psidium*, several species). *Myrtaceæ.*

The guavas grow readily from seeds, and plants will often bloom when a year and a-half old. They may also be multiplied by layers, and by cuttings either under glass or in the open.

Guelder Rose. See Viburnum.

Guernsey Lily. See Nerine.

Gumbo. See Okra.

Gum, Sweet. See Liquidambar.

Gum-tree. See Eucalyptus.

Gunnera. *Halorageæ.*

Propagated by division. It is very difficult to raise from seed.

Gymnocladus (Kentucky Coffee-tree). *Leguminosæ.*

Readily increased by seeds, which start better if soaked for a few hours in hot water. Also by root-cuttings.

Gymnogramme. See Ferns.

Gymnyostachys. *Aroideæ.*

Propagation is effected by suckers and divisions.

Gynerium (Pampas Grass). *Gramineæ.*

Seeds, under glass in the north. Also increased by dividing the tufts.

Gynura. *Compositæ.*

Increased easily by cuttings and seeds.

Gypsophila. *Caryophyllaceæ.*

Propagated by seeds, division or cuttings.

Habrothamnus. See Cestrum.

Hackberry. See Celtis.

Hæmanthus (Blood Flower). *Amaryllidaceæ.*

Bulbels, which should be removed and potted when the plants are commencing new growth, and be kept in a close pit or house till established. Seeds are rarely used.

Hæmodorum (Australian Bloodroot). *Hæmodoraceæ.*

Increased by dividing the roots in spring.

Hakea. *Proteaceæ.*

Well-ripened cuttings, placed in sandy peat under glass, in a cool house. Seeds, when obtainable, can be used.

Halesia (Silver-bell, or Snowdrop Tree). *Styracaceæ.*

Seeds, which rarely germinate till the second year. They should be stratified or kept constantly moist. Propagation is also effected by layers, or by cuttings of the roots in spring and autumn. Layers are commonly employed in this country.

Halimodendron (Salt-tree). *Leguminosæ.*

Freely increased by seeds, layers or cuttings. May also be grafted on common laburnum.

Hamamelis (Witch-hazel). *Hamamelideæ.*

All grow from seeds or layers, and the Japan species succeed if grafted on American species in the greenhouse. *H. Virginica* may be readily propagated by layers.

Hamelia. *Rubiaceæ.*

Seeds. Cuttings which are nearly ripe will root during the early part of summer under glass, with heat.

Hamiltonia. *Rubiaceæ.*

Seeds. Half-ripened cuttings, placed in sand under glass.

Hardenbergia. *Leguminosæ.*

Seeds may be used; also increased by division. Cuttings, made of the firm young side shoots in spring, will grow if inserted under a bell-glass, and placed in a warm frame or pit, without bottom heat.

Harebell. See Campanula.

Harpalium. See Helianthus.

Haw, Hawthorn. See Cratægus.

Hazel. See Corylus.

Heartsease. See Viola.

Heath. See Erica.

Heather. See Calluna; also Erica

Hedera (Ivy). *Araliaceæ.*

Seeds. Layers. The rooted portions of the vine may be severed and treated as independent plants. Cuttings may be made in autumn from any firm shoots, and inserted in pots or in the open ground. If they are placed in heat and kept shaded until roots are formed, good plants are obtained much sooner than when placed in a coldframe or in the open air. Named varieties are grafted on the stock of any common strong climbing form.

Hedychium (Indian Garland Flower). *Scitamineæ.*

Seeds, rarely. Increased by dividing the rhizomes in spring, when the plants are repotted.

Hedysarum. *Leguminosæ.*

Propagation is effected by means of seeds and division.

Helenium. *Compositæ.*

Increased by seeds or divisions.

Helianthemum (Rock-Rose, Sun-Rose). *Cistaceæ.*

The annuals are raised from seeds. The perennials may also be raised from seeds, but it is better to trust to layers and to cuttings, which will root freely in a sandy soil, if kept shaded until established.

Helianthus, including Harpalium (Sunflower). *Compositæ.*

By seeds, which may be sown in pots, and the seedlings transferred, or in the open ground in spring. Also divisions. Of perennial sorts, use the underground shoots or "creepers," treating as for cuttings.

Helichrysum, Elichrysum (Everlastings). *Compositæ.*

The annual species and the varieties of *H. bracteatum* may be raised from seed, sown in a light heat in early spring, and afterwards transplanted; or sown in the open ground a little later. The perennial species are increased by cuttings in spring, in a close frame without heat.

Heliconia. *Scitamineæ.*

May be increased by seeds, but the best method is by division of the rootstock in spring when growth commences. Separate pieces may be placed in pots, and grown in a moist stove temperature, repotting when necessary; or they may be planted out in the stove, if desired.

Heliotropium (Heliotrope). *Borraginaceæ.*

Seeds. The common practice is to use cuttings. These can be taken at almost any season, if good growing shoots are to be had. They start readily in sand or soil on a cutting bench, or under a frame. Plants for bedding are struck in late winter from stocks which are in a vigorous condition.

Helipterum, including Rhodanthe. *Composilæ.*

Seeds may be sown in early spring, under cover.

Helleborus (Black Hellebore, Christmas Rose). *Ranunculaceæ.*

Seeds may be sown as soon as ripe. Strong and healthy root divisions are also employed. See, also, Veratrum.

Helonias. *Liliaceæ.*

Propagation is effected by seeds, and slowly by root divisions.

Hemerocallis (Day Lily). *Liliaceæ.*

Increased by divisions. *H. Middendorfii* and some others by seeds.

Hemlock Spruce (*Tsuga Canadensis*). *Coniferæ.*

Seeds. Named varieties top-worked on seedlings. Handled the same as Abies and Picea, which see.

Hemp. See Cannabis.

Hepatica. *Ranunculaceæ.*

Can be propagated by division; also by seeds.

Heracleum (Cow Parsley, Cow Parsnip). *Umbelliferæ.*

Readily increased by seeds or divisions.

Herbertia. *Iridaceæ.*

Propagated by means of seeds or bulbels.

Hesperis (Dame's Violet, Rocket). *Cruciferæ.*

The single sorts are increased by seeds; the double forms by carefully dividing the roots, or by cuttings.

Heuchera (Alum Root). *Saxifragaceæ.*

Seeds. Readily increased by dividing the crowns during spring.

Hibiscus. *Malvaceæ.*

Seeds. Also by divisions and layers. Cuttings of green wood are commonly used, made in summer for hardy species or in early spring for tender ones. Cuttings of ripened wood may be taken in fall, and stored until spring in a rather dry place. The variegated sorts do better if grafted upon strong stocks.

Hicoria, Carya (Hickory, Pecan, etc.). *Juglandaceæ.*

Increased chiefly by seeds, which should be stratified or planted (about 3 inches deep) as soon as ripe; also by root-sprouts. Seeds are sometimes planted at intervals in the field where the trees are to stand; but this practice is not to be recommended. Cuttings of the ends of growing roots are often successful. The hickory can be grafted. Best results are probably obtained by veneer or splice-grafting in winter, on potted stocks. Cleft-grafting can be employed outdoors, however, the stub being cut 3 to 6 inches below the ground, and the cions covered with earth, as for grafting the grape. The cions must be perfectly dormant, and are safer, therefore, if they have been kept on ice or in a very cold cellar. Saddle-grafting upon young twigs is sometimes used. Shield- and flute-budding often succeed in the hickories, as, in fact, many kinds of graftage do; but the skill of the operator is more important than the method. See also Pecan.

Hippeastrum (Equestrian Star). *Amaryllidaceæ.*

Seeds may be sown as soon as ripe in well-drained pots or pans of sandy loam, slightly covered, and placed in a temperature of about 65°. For increasing by divisions—which is the usual way—the old bulbs should be taken from the pots and carefully separated, with the least possible injury to the roots. This should be done when the plants are at rest, and the offsets should be placed singly in pots. Keep the bulb about two-thirds above the level of the soil, dispose the roots evenly, and plunge in bottom heat, in a position exposed to the light.

Hippophae (Sallow Thorn, Sea Buckthorn). *Elæagnaceæ.*

May be increased by seeds, suckers, layers, and cuttings of the roots.

Hoffmannia, Higginsia. *Rubiaceæ.*

Insert cuttings in sandy soil under cover, in bottom heat.

Hog Plum. See Spondias.

Holly. See Ilex.

Hollyhock (Althæa). *Malvaceæ.*

Seeds should be sown as soon as ripe—in summer—in pots or pans, and placed in a slight bottom heat or in the open air. In either case, place the seedlings in 3-inch pots, and winter in a coldframe. Dividing the roots, after flowering is over, by separating the crown, so as to preserve one or more buds and as many roots as possible to each piece. Cuttings of young shoots 3 inches long, taken off close to the old root at nearly the same time, should be placed singly in small pots of light, sandy soil and kept close, and shaded in a coldframe until rooted. If cuttings are made during winter, a gentle bottom heat must be given. Also grafted (see page 129). See also Althæa.

Honesty. See Lunaria.

Honey Locust. See Gleditschia.

Honeysuckle. See Lonicera.

Hop. See Humulus.

Hop Hornbeam. See Ostrya.

Horehound (*Marrubium vulgare*). *Labiatæ.*

Seeds, in early spring. Division.

Horkelia. See Potentilla.

Hornbeam. See Carpinus.

Horse-Chestnut. See Æsculus.

Horse-Radish (*Nasturtium Armoracia*). *Cruciferæ.*

Root cuttings ("sets"). These are made from the small side roots when the horse-radish is dug. They may be anywhere from one-fourth to one inch in diameter, and 3 to 6 inches long, one end being cut slanting, to mark it. These are planted obliquely, 2 to 4 inches deep, in spring. They may be buried during winter. (Fig. 64.) The old crowns may be planted, but they make poorer roots.

Hottonia. *Primulaceæ.*

Propagation is effected by seeds, and divisions in spring.

House-Leek. See Sempervivum.

House Plants. The common conservatory plants, like fuchsias, geraniums, carnations, and the like, give best results when allowed to bloom but one year. They are then thrown away and their places supplied by other plants. Cuttings are generally made in late winter or spring for the next winter's bloom. These cuttings are slips (page 65) of the growing wood. See the various species, under their respective heads.

Houstonia. *Rubiaceæ.*

Seeds. May also be increased by carefully made divisions in autumn or spring.

Hovea, Poiretia. *Leguminosæ.*

Propagation is best effected by seeds, sown in well-drained pots of sandy peat soil in spring, and placed in a gentle bottom heat. Cuttings are difficult to strike.

Hovenia. *Rhamnaceæ.*

Increased by seeds. Root cuttings are also used.—Ripened cuttings should be placed in sand, under a hand-glass.

Hoya (Honey Plant, Wax Flower). *Asclepiadaceæ.*

For layering, good-sized shoots should have a few of their leaves removed, and should then be put in pots of soil until rooted. The plants may afterwards be grown on, and repotted according to their strength. Cuttings may be taken in spring or later in the year, from shoots of the preceding summer's growth, and placed in a compost of peat and sand, and plunged in bottom heat in a frame. A slight shade and careful watering will be necessary. *H. bella* does best when grafted on a stronger growing sort.

Huckleberry. See Vaccinium and Whortleberry.

Humulus (Hop). *Urticaceæ.*

It may be propagated by seeds, or by divisions in spring. Ordinarily, however, the species is increased by hardwood cuttings of two-bud lengths from the best old shoots, and made in spring. Leave the top bud just above the ground.

Hyacinthus (Hyacinth). *Liliaceæ.*

Seeds are employed for the production of new varieties. These are sown the same season they mature, in light, sandy soil, and are covered not more than a half-inch deep. In four or five years, or sometimes even longer, the bulbs will be large enough to flower. Varieties are perpetuated by means of the bulbels, which form freely upon some varieties. These are treated in much the same manner as mature bulbs, or they may be handled in pans or flats. They make flower bulbs in two or three years. To increase the numbers of these bulbels, the bulbs are variously cut by the Dutch growers. These practices are described and illustrated on pages 28 and 29, Figs. 21–23. Hyacinths can be propagated by leaf cuttings. Strong leaves should be taken in early spring and cut into two or three portions, each portion being inserted about an inch in good sandy loam, and given a temperature of about 75°. In eight or ten weeks a bulblet will form at the base of the cutting (see page 60). The lower leaves give better results than the upper ones. These bulblets are then treated in the same manner as bulbels. For *Hyacinthus candicans*, see Galtonia.

Hydrangea, Hortensia. *Saxifragaceæ.*

The hardy species are usually propagated by green cuttings in summer, under glass (see Fig. 75). The tender ones (*H. Hortensia*, the var. Otaksa, etc.) are increased by cuttings taken at any time from vigorous young wood, usually in late winter. Layers are occasionally employed, and suckers can be separated from some species. Sometimes the hardy species are forced for purposes of propagation by cuttage. *H. quercifolia* is propagated by little suckers or "root pips." *H. paniculata grandiflora* can easily be propagated from the young wood, taken in June and planted under glass.

Hymenocallis. *Amaryllidaceæ.*

Treated the same as Pancratium, which see.

Hypericum. *Hypericaceæ.*

Easily increased by seeds, cuttings, or by strong pieces of the roots of creeping-rooted species. Hard-wooded cuttings, taken in fall, are commonly used.

Hypoxis. *Amaryllidaceæ.*

Propagation is effected by seeds and offsets.

Hyssop (*Hyssopus officinalis*). *Labiatæ.*

Seeds. Division.

Iberis (Candytuft). *Cruciferæ.*

The annuals and biennials are increased by seeds sown in light sandy soil, in spring or autumn. The sub-shrubby sorts are also increased by seeds sown in spring, but more often by divisions or by cuttings.

Ilex, including Prinos (Holly). *Ilicineæ.*

Seeds, which should be stratified. They are often cleaned of the pulpy coat by maceration. The seeds rarely germinate until the second year. Varieties are perpetuated by graftage. The veneer-graft, upon potted plants, is usually employed, but other methods may be successful. Budding is sometimes performed.

Illicium (Aniseed-tree). *Magnoliaceæ.*

Seeds. Cuttings of young ripened shoots may be made during summer and should be placed in sandy soil, under a glass.

Imantophyllum. *Amaryllidaceæ.*

Seeds. Usually increased by division or by means of bulbels.

Impatiens. See Balsam.

Indian Corn. See Maize.

Indian Fig. See Opuntia.

Indian Shot. See Canna.

Indigofera (Indigo). *Leguminosæ.*

Propagated by seeds. Cuttings of young shoots may be inserted in sandy or peaty soil under glass, in slight heat.

Inula, Elecampane. *Compositæ.*

Readily increased by seeds or by division.

Ionidium, Solea. *Violaceæ.*

The herbaceous species are increased by seeds and by divisions. The shrubby sorts are increased by cuttings, which will root in sand, in a frame.

Ipomœa, including Quamoclit (Moonflower, Morning Glory). *Convolvulaceæ.*

All the annual species are grown from seeds. Seeds of moonflowers should usually be filed or cut on the point, and started in a rather high temperature. The perennials are also increased by seedage, but they may be raised from cuttings struck in a forcing-house or a frame. The moonflowers often do better in the north from cuttings than from seeds. *I. Horsfalliæ* is largely propagated by layers, and other species may be treated in the same way. Division is sometimes employed. *I. pandurata* can be propagated by root cuttings. Also grafted (see page 129).

Ipomopsis. See Gilia.

Iresine, Achyranthes. *Amarantaceæ.*

Seeds rarely. Increased readily by cuttings. For summer bedding in the north, cuttings should be started in February or March. For use as window plants, they should be taken in late summer.

Iris, including Xiphion (Blue Flag). *Iridaceæ.*

Seeds grow readily and give good results, and they are usually produced freely, especially in the bulbous species. Sow as soon as ripe in light soil in some protected place. The bulbous species produce bulbels, which may be used for multiplication. The rhizomatous species are propagated by dividing the rhizome into short-rooted pieces. Or when the rhizomes lie on the surface of the ground and do not root readily, they may be layered.

Isonandra (Gutta-Percha Tree). *Sapotaceæ.*

Insert cuttings in sandy soil, under glass, in heat.

Itea. *Saxifragaceæ.*

Propagated by seeds or by suckers, in spring; and in autumn by layers.

Ivy. See Hedera and Ampelopsis.

Ixia. *Iridaceæ.*

Seeds may be sown in pans of sandy soil in autumn, and placed in a cool frame. Propagation by bulbels is much quicker, and is the usual method.

Ixiolirion. *Amaryllidaceæ.*

Increased by seeds, and by bulbels.

Ixora. *Rubiaceæ*

Seeds. Usually increased by short-jointed green cuttings placed in a close frame with a strong bottom heat.

Jacaranda. *Bignoniaceæ.*

Cuttings of half-ripened shoots may be made in early summer and placed in sand over sandy peat, in heat, and kept shaded. Also seeds.

Jacobæan Lily. See Amaryllis.

Jacobinia. See Justicia.

Jasminum (Jasmine, Jessamine). *Oleaceæ.*

Sometimes by seeds, but usually by cuttings of the nearly ripened wood, under glass. Cuttings of ripe wood are also employed, and layers are often used.

Jatropha. *Euphorbiaceæ.*

Cuttings made of firm young shoots will strike in sandy soil in a strong bottom heat. The cuttings, if very fleshy, may be dried a few days before setting them.

Jeffersonia. *Berberidaceæ.*

Seeds should be sown as soon as ripe, or divisions may be made.

Jerusalem Artichoke (Girasole). See Artichoke.

Jessamine, Yellow. See Gelsemium.

Jonquil. See Narcissus.

Jubæa (Coquito Palm of Chili). *Palmaceæ.*

Propagation is effected by seeds.

Judas-tree. See Cercis.

Juglans (Walnut and Butternut). *Juglandaceæ.*

All the species are readily propagated by means of stratified nuts. Do not allow the nuts to become dry. Artificial cracking should not be done. In stiff soils the seedlings are apt to produce a long tap-root which renders transplanting difficult after the first year or two. The tap-root may be cut by a long knife while the tree is growing, or the young seedling may be transplanted. Particular varieties are perpetuated by grafting or budding with any of the common methods; but the skill of the grafter is more important than the method. In the north, they are sometimes worked indoors in pots. Common shield-budding works well, if the sap is flowing freely in the stock. Flute-budding is often employed. The improved native sorts are root-grafted in winter. Old trees can be top-grafted like apple trees (see page 123). If nursery stocks are grafted, it is usually best to insert the cions below ground, as for grapes. In all walnut grafting, it is generally preferred that only one scarf or cut of the cion should traverse the pith. It is very important that the cions be kept perfectly dormant.

The "English" walnut (*J. regia*) is mostly grown direct from seed in this country, and the different varieties usually come true. In California, the native walnut (*J. Californica*) is often used as a stock for this species, and flute-budding on branches a half-inch or more in diameter is often practised. Twig- or prong-budding (Fig. 96) is sometimes employed.

Jujube (*Zizyphus Jujube*). *Rhamnaceæ.*

Seeds and cuttings.

Juncus (Rush, Bulrush). *Juncaceæ.*

Seeds. The perennials may be increased by division. *Scirpus Tabernæmontanus variegatus* of florists is a form of *Juncus effusus*, and is increased by division of the stools.

Juneberry (*Amelanchier oblongifolia*). *Rosaceæ.*

Increased by using the sprouts which form freely about the old plants; also by seeds. The cultivated dwarf Juneberry is multiplied by suckers. See Amelanchier.

Juniperus (Juniper, Red Cedar, Savin). *Coniferæ.*

Increased readily by seeds, which, however, often lie dormant until the second year. Red cedar seed is one of the species which lie dormant a year. They germinate more readily if the pulp is removed by maceration or by soaking with ashes for a few days. Green cuttings, in sand under glass, root easily; or mature cuttings may be taken in fall and placed in a coldframe, in which they will need little protection during winter. Some varieties require a long time to root. Most of the named varieties may be grafted on imported Irish stocks, which are much used in some parts of the country. They may be veneer-grafted and handled in a cool house.

Justicia, including Jacobinia and Sericographis. *Acanthaceæ.*

Seeds occasionally. The species strike readily from short green cuttings on a cutting-bench or under a frame.

Kadsura, Sarcocarpon. *Magnoliaceæ.*

Seeds. Cuttings, made of nearly ripened shoots, which should be placed in sand under glass.

Kaki. See Persimmon.

Kalanchoe. *Crassulaceæ.*

Propagated by seed, but cuttings, when obtainable, are better.

Kale (*Brassica oleracea*, vars.). *Cruciferæ.*

By seeds, sown in the open in spring in the north, or in the fall in the south.

Kalmia (Mountain Laurel, Calico-bush). *Ericaceæ.*

May be increased by seeds, which should be sown in shallow pans of sandy peat or sphagnum, and kept in a coldframe until the seedlings are large enough to transfer to the open air after being hardened off. By cuttings of young shoots in sandy peat, placed in a shady situation under a hand-glass (with much difficulty). Also by layers. Usually obtained from the woods. Varieties are veneer-grafted under glass, upon unnamed stocks.

Kennedya. *Leguminosæ.*

Seeds may be sown in spring or summer, or cuttings of rather firm side shoots may be made at the same time, and placed in peaty soil, in a close, warm frame.

Kentia. *Palmaceæ.*

Increased by seeds, placed in light, sandy soil, with heat.

Kentucky Coffee-tree. See Gymnocladus.

Kerria. *Rosaceæ.*

Propagated by divisions, layers, and by cuttings of young shoots, inserted under a hand-light, or by ripened cuttings. In this country, oftener increased by ripe wood in fall.

Kleinhovia. *Sterculiaceæ.*

Seeds. Make cuttings of the young ripened shoots, and place in sand, in heat, under glass.

Klugia. *Gesneraceæ.*

Seeds. Propagated usually by cuttings.

Knightia. *Proteaceæ.*

Make cuttings of ripened shoots with upper leaves on, and place in sandy soil under glass, in a very gentle bottom heat.

Kniphofia, Tritoma. *Liliaceæ.*

Increased by seeds, or by divisions of the crown in early spring.

Kœlreuteria. *Sapindaceæ.*

Propagated in spring by seed, by layers in autumn, and by cuttings of the young shoots in spring; also by root-cuttings.

Krameria. *Polygalaceæ.*

Cuttings, set in sand under glass, in spring. Seeds.

Kumquat (*Citrus Japonica*). *Rutaceæ.*

Worked on stocks of orange (which see).

Laburnum (Golden Chain). *Leguminosæ.*

The species may be increased by seeds. Layers and suckers are often used. The varieties by grafting or budding on the common sorts. See Cytisus.

Lachenalia. *Liliaceæ.*

Seeds. Bulbels.

Lælia. *Orchidaceæ.*

Increased by pseudo-bulbs, as in cattleya. See also under Orchids.

Lagerstrœmia (Crape Myrtle). *Lythraceæ.*

Seeds. Layers. Cuttings of firm, small side shoots may be made in spring, and placed in bottom heat.

Lagetta (Lace Bark). *Thymelæaceæ.*

Usually increased by cuttings of firm shoots, placed in sand under glass, in bottom heat.

Lantana. *Verbenaceæ.*

Seeds, which give new varieties. Cuttings, in fall or spring, from good growing wood, in sand in a warm house or frame.

Lapageria, Phænocodon. *Liliaceæ.*

Sow seeds as soon as ripe in a sandy peat soil, and keep in a moderate heat. Increased by layers of firm, strong shoots.

Larix (Larch, Tamarack). *Coniferæ.*

Seeds should be kept dry over winter and planted early in spring. Shade the young plants. Varieties, as the weeping sorts, are worked upon common stocks. The grafting may be done by the whip method, outdoors early in spring. Rare sorts are sometimes veneer-grafted under glass.

Larkspur. See Delphinium.

Lasiandra. *Melastomaceæ.*

Propagated by cuttings of the growing wood under glass.

Lasiopetalum. *Sterculiaceæ.*

Seeds. Make cuttings in spring of the half-ripened wood, and insert in sand, under glass.

Latania. See Livistona.

Lathyrus (Sweet Pea, Vetchling). *Leguminosæ.*

Seeds, sown very early in the open. The perennials also by seeds, sometimes by division. The sweet pea may be sown before frosty weather is passed, and south of Norfolk it is usually satisfactory if sown in the fall. The everlasting pea (*L. latifolius*) is increased by seeds, division and cuttings.

Lattice-leaf. See Ouvirandra.

Laurel, Mountain. See Kalmia.

Laurus (Laurel). *Lauraceæ.*

Increased by seeds, layers, and by cuttings, placed under a hand-glass in sandy soil. Also propagated by root-cuttings.

Laurestinus. See Viburnum.

Lavandula (Lavender). *Labiatæ.*

Divisions. When the flowers are fully expanded, cuttings may be made. These should be inserted in sandy soil, under a frame.

Lawsonia. *Lythraceæ.*

Increased by cuttings of ripened shoots, placed in sand under a glass, in heat.

Layia. *Compositæ.*

Increased by seeds, sown in a hotbed, or in the open border in the south.

Leaf-Beet, or Chard. See Beet.

Ledum (Labrador Tea). *Ericaceæ.*

Propagated by seeds and divisions, but principally by layers, in sandy peat soil.

Leek (*Allium Porrum*). *Liliaceæ*.

Seeds, sown very early in the spring, either outdoors or in a coldframe.

Leiophyllum (Sand Myrtle). *Ericaceæ*.

May be freely increased by seeds, sown in pans and placed in a frame. By layers in autumn.

Lemon (*Citrus Medica*, var. *Limon*). *Rutaceæ*.

The named sorts are budded upon either orange or lemon stocks. Orange stocks are probably most generally preferred, as they are adapted to a great variety of soils, and vigorous trees nearly always result. The budding is performed in the same manner as upon the Orange, which see. Lemons are often grown from cuttings of the mature wood, which are set in the open ground as soon as the spring becomes warm, or in a frame. Stocks for budding upon are sometimes grown from cuttings in this way.

Lentil (*Ervum*, various species). *Leguminosæ*.

Seeds, sown in early spring.

Leonotis (Lion's Ear, Lion's Tail). *Labiatæ*.

Seeds. Increased by cuttings, which root freely in a gentle bottom heat, in early spring.

Leontice. *Berberidaceæ*.

May be increased by seeds or by suckers.

Leontopodium (Edelweiss, Lion's Foot). *Compositæ*.

May annually be raised from seeds, or the old plants may be divided in spring. The seeds must be kept in a dry place throughout the winter.

Lepachys. See Rudbeckia.

Leptosyne. *Compositæ*.

Propagated by seeds.

Lessertia. *Leguminosæ*.

Propagation by seeds, or by divisions in spring.

Lettuce (*Lactuca sativa*). *Compositæ*.

Seeds, which may be sown under glass or in the open. In the middle and southern states, the seeds may be sown in the fall, and the plants protected during cold by a mulch; or the plants may grow during winter in the warmer countries.

Leucoium (Snowflake). *Amaryllidaceæ.*

Seeds, for producing new sorts. Propagation is commonly effected by bulbels, which should be secured as soon as possible after the foliage ripens.

Leucothoë. *Ericaceæ.*

Increased by seeds, which should be covered very lightly. By divisions of established plants in autumn or winter. Also by layers.

Lewisia. *Portulacaceæ.*

Propagated by seeds, or by divisions in spring.

Liatris (Blazing Star, Button Snake-root). *Compositæ.*

Seeds are usually sown early in autumn. Divisions may be made in spring.

Libonia. *Acanthaceæ.*

Seeds are rarely employed. Usually increased by short green cuttings, like fuchsia and pelargonium.

Licuala. *Palmaceæ.*

Seeds may be sown in spring in a sandy soil, and placed in a strong, moist bottom heat.

Ligustrum (Privet, Prim). *Oleaceæ.*

Stratified seeds. Division. The named varieties are grown from cuttings, either of green or ripe wood.

Lilac. See Syringa.

Lilium (Lily). *Liliaceæ.*

Seeds — giving new varieties in the variable species — should be sown as soon as ripe in well-drained pans of sandy peat, slightly covered with similar soil and a layer of moss, and placed in a cool frame. Usually increased by bulbels, which should be planted a few inches apart in prepared beds. Sometimes small bulblets form in the axils of the leaves, and these are used in the same manner as bulbels. Bulb-scales are often employed for the multiplication of scarce kinds. Those which produce large and loose bulbs, as *L. candidum*, may be increased by simple division. These operations are described on pages 26 to 31.

Lily-of-the-Valley. See Convallaria.

Lime (*Citrus Limetta* and *C. Medica*, var. *acris*). *Rutaceæ.*

Seeds, which usually reproduce the variety. Some varieties are budded upon strong seedlings.

Lime-tree. See Tilia.

Limnocharis. *Alismaceæ.*

Increased by seeds, by divisions, and by runners.

Linaria (Toadflax). *Scrophulariaceæ.*

Increased by seeds sown in light soil, in early spring. Or by divisions made in spring or autumn. The greenhouse species are ordinarily grown from seeds, which should be carefully sown in finely pulverized soil. Cuttings may also be used.

Linden. See Tilia.

Lindera (Spice-bush, Benzoin). *Lauraceæ.*

Seeds, stratified. Divisions. Cuttings, as in Cornus.

Lindleya. *Rosaceæ.*

Increased by ripened cuttings under glass in bottom heat; or by grafting on the hawthorn.

Linnæa. *Caprifoliaceæ.*

Naturally increased by layers or runners. Seeds are rarely employed

Linum (Flax). *Linaceæ.*

Propagated by seeds, the hardy species sown outdoors and the tender ones under glass. Cuttings may be taken from firm shoots and inserted in a sandy position under glass. The ordinary flax is sown directly in the field. See Reinwardtia.

Lippia. *Verbenaceæ.*

Seeds. Usually by cuttings of young shoots, which will root freely in sandy soil in a close, warm frame. If it is not possible to secure the necessary heat, cuttings of the hard wood can be used in autumn, under glass.

Liquidambar Sweet Gum). *Hamamelideæ.*

Seeds, which should be stratified or sown as soon as ripe. Many of the seeds may lie dormant until the second year.

Liquorice. See Glycyrrhiza.

Liriodendron (Tulip-tree, Whitewood). *Magnoliaceæ.*

Increased by seeds, which are stratified as soon as they are ripe, and sown the following spring. Named varieties are grafted on seedlings. The seeds of the tulip-tree are apt to be hollow, especially those grown along the eastern limits of the distribution of the species.

Litchi. See Nephelium.

Livistona, Latania (Fan Palm). *Palmaceæ.*

Seeds, sown in a sandy soil and placed in a gentle bottom heat.

Lloydia. *Liliaceæ.*

Seeds rarely. Increased by bulbels, or by the creeping shoots, leaving a bulb at the extremity.

Loasa, including Illairea. *Loasaceæ.*

All are easily increased by seeds sown in a light, sandy soil, usually under cover. Cuttings are rarely used.

Lobelia. *Lobeliaceæ.*

Ordinarily increased by seeds, which are more certain if handled in pans or flats under glass. Cuttings from vigorous shoots may be employed, and strong plants of some species may be divided. The cardinal flower (*L. cardinalis*) is grown from seeds carefully sown in fine soil, usually under cover.

Locust-tree. See Robinia ; also Gleditschia.

Loddigesia. *Leguminosæ.*

Increased in spring by cuttings placed under glass, in sandy soil.

Lœselia. *Polemoniaceæ.*

Seeds. Cuttings of half-ripened shoots in sand under glass.

Logania. *Loganiaceæ.*

Propagated by cuttings of firm side shoots inserted in sandy soil, under glass.

Loiseleuria. *Ericaceæ.*

Propagation by layers ; very rarely by seeds, which are slow and uncertain. Like Andromeda.

Lonicera, including Caprifolium and Xylosteum (Honeysuckle, Woodbine). *Caprifoliaceæ.*

Seeds, for new varieties. Sow as soon as ripe, or stratify, first removing them from the pulp. The upright species are commonly grown from layers and from cuttings of dormant wood. The creepers are mostly grown from dormant cuttings.

Lophospermum. See Maurandia.

Loquat See Photinia.

Lotus. *Leguminosæ.*

The species may be raised annually from seeds. Increased also by cuttings.

Lotus of the Nile. See Nymphæa ; also Nelumbo.

Lovage (*Levisticum officinale*). *Umbelliferæ.*

Seeds sown in the open ground, and division.

Lucerne. See Medicago.

Luculia. *Rubiaceæ.*

Sow seeds in sandy soil and place in a little heat. Cuttings of young shoots may be inserted in spring, under glass, in gentle bottom heat for the first two or three weeks. Insert immediately after cutting, and water freely.

Luffa (Dish-cloth Gourd). *Cucurbitaceæ.*

Seed, sown in the open, or in the north better started in pots in early spring.

Lunaria (Honesty). *Cruciferæ.*

Propagated by seeds or by division.

Lupinus (Lupine). *Leguminosæ.*

Seeds of annuals may be sown in the open border during early spring. The perennials may be increased the same way, or by dividing the stronger-growing plants during very early spring.

Lycaste. *Orchidaceæ.*

Division and pseudo-bulbs. (See also under Orchids.)

Lychnis, including Agrostemma, Viscaria. *Caryophyllaceæ.*

Increased readily in spring by seeds, division or cuttings.

Lycium (Matrimony Vine, Box Thorn). *Solanaceæ.*

Increased by seeds, suckers, layers; and by cuttings made in autumn or spring.

Lycopodium (Club-moss). *Lycopodiaceæ.*

Spores, as for Ferns (which see). Short cuttings in pans or pots.

Lygodium (Hartford Fern, Climbing Fern). *Filices.*

By spores, and divisions of the root. See Ferns.

Lyonia. *Ericaceæ.*

Increased by seeds, which should be sown very carefully in sandy peat soil. Also by layers.

Lysimachia (Loosestrife). *Primulaceæ.*

Propagation is easily effected by seeds; by divisions in late autumn or early spring; and by cuttings.

Lythrum (Loosestrife). *Lythraceæ.*

Seeds and divisions are the usual methods. Cuttings are employed for some species.

Maclura (Osage Orange). *Urticaceæ.*

Sow seed in the spring. Soak in warm water a few days before sowing.

Madeira Vine. See Boussingaultia.

Magnolia. *Magnoliaceæ.*

Seeds are commonly used. The coverings should be macerated in the very pulpy species. The cucumber trees and some others are sown directly in autumn. The seeds of any species should not be allowed to become thoroughly dry. Magnolias strike well from green cuttings, cut to a heel and handled under glass. Layers are often used. Named varieties are veneer- or side-grafted upon strong stocks. The cucumber tree (*M. acuminata*) is used as a stock for all species. The umbrella tree (*M. Umbrella*) is also a good stock.

Mahernia. *Sterculiaceæ.*

Propagated during summer by cuttings of young shoots, 1 or 2 inches long, inserted in sandy soil under glass.

Mahonia. See Berberis.

Maidenhair-tree. See Ginkgo.

Maize, Indian Corn (*Zea Mays*). *Gramineæ.*

Seeds (properly fruits), planted upon the approach of warm weather.

Malcolmia. *Cruciferæ.*

Propagated by seeds.

Mallow. See Malva.

Malope. *Malvaceæ.*

Seeds may be sown either under glass in early spring, or in the open border a month or two later.

Malpighia. *Malpighiaceæ.*

Cuttings of nearly ripened shoots may be made in summer (with leaves), or under glass.

Malva (Mallow). *Malvaceæ.*

The annuals by seeds only. The perennials may be increased by seeds, divisions or cuttings.

Malvaviscus. *Malvaceæ.*

Increased by seeds, and by cuttings of side shoots, placed under glass, in heat.

Mammea (Mammee Apple, St. Domingo Apricot). *Guttiferæ.*

Seeds. Cuttings of half-ripened shoots should be taken with the leaves on and placed in a frame.

Mammillaria. See Cactus.

Mandevilla. *Apocynaceæ.*

Propagated by seeds, layers, or cuttings of half-ripened wood.

Mandiocca. See Manihot.

Mandragora (Mandrake). *Solanaceæ.*

Propagated by seeds or by divisions.

Mandrake. See Mandragora and Podophyllum.

Manettia. *Rubiaceæ.*

Seeds are sometimes employed. Usually increased by cuttings of young shoots. Root-cuttings are sometimes made.

Mangifera. See Mango.

Mango (*Mangifera Indica*). *Anacardiaceæ.*

Stocks are obtained by seeds. The seeds usually have more than one embryo, sometimes as many as ten. Each embryo will produce a distinct plant. The embryos may be separated before planting, but it is preferable to separate the young plantlets soon after germination, before they grow together, as they are apt to do. The seeds germinate better if the hard shell is removed before planting. Seeds retain their vitality but a few days, and if to be shipped for sowing they should be enclosed in wax. Seedlings begin to bear from the third to the sixth years. Varieties are inarched upon other stocks.

Mangostana, Mangosteen. See Garcinia.

Manicaria, Pilophora. *Palmaceæ.*

Increased by seeds, which should be sown in a strong, moist heat.

Manihot, Janipha, Mandiocca. *Euphorbiaceæ.*

Propagation is effected by cuttings of young and rather firm shoots, placed in sandy peat under glass, in bottom heat. For the propagation of *M. Aipe*, see Cassava.

Mantisia. *Scitamineæ.*

Propagated usually by divisions, made just as growth commences.

Maple. See Acer.

Maranta. See Calathea.

Marguerite, or **Paris Daisy** (*Chrysanthemum frutescens* and *C. fœniculaceum*). *Compositæ.*

Cuttings, as described for Chrysanthemum, which see.

Marigold. See Tagetes and Calendula.

Mariposa Lily. See Calochortus.

Marsdenia. *Asclepiadaceæ.*

In spring, cuttings may be made and inserted in sand, under glass.

Marsh-Mallow. See Althæa.

Marsh-Marigold. See Caltha.

Martynia (Unicorn Plant). *Pedalineæ.*

Seeds, sown where the plants are to grow, or started under glass in the north.

Masdevallia. *Orchidaceæ.*

Division. See also under Orchids.

Matthiola (Stock, Gilliflower). *Cruciferæ.*

Seeds, sown either under cover or in the garden. Grows readily from cuttings.

Maurandia, including Lophospermum. *Scrophulariaceæ.*

Seeds, sown in heat. Cuttings of young growth under glass.

Maxillaria. *Orchidaceæ.*

Division of the plants, and also of the pseudo-bulbs. See also under Orchids.

May-Apple. See Podophyllum.

Meadow-Rue. See Thalictrum.

Meadow-Sweet. See Spiræa.

Meconopsis. *Papaveraceæ.*

Seeds, sown in early spring in a gentle heat. Also propagated by division.

Medicago (Lucerne, Medick). *Leguminosæ.*

Propagated by seeds or by division. Alfalfa (*M. sativa*) by seeds in spring.

Medinilla. *Melastomaceæ.*

Cuttings of young wood in strong, close heat.

Medlar (*Pyrus* [or *Mespilus*] *Germanica*). *Rosaceæ.*

Stocks are grown from stratified medlar seeds, and the plant may be worked upon these, the thorn, or the quince.

Megarrhiza. *Cucurbitaceæ.*

Propagation is effected by seeds, sown in gentle heat in spring.

Melaleuca. *Myrtaceæ.*

Seeds. In spring, cuttings getting firm at the base may be made about 3 inches in length. Place in a compost of peat and sandy loam.

Melastoma. *Melastomaceæ.*

Make cuttings during spring, and place in sandy peat under glass, in heat.

Melia (Bead-tree, Pride of India). *Meliaceæ.*

Seeds, sown as soon as ripe. Cuttings of growing wood under glass.

Melicocca (Ginep, Spanish Lime). *Sapindaceæ.*

Seeds. Place ripened cuttings in sand under glass, in heat.

Melocactus. See Cactus.

Melon (*Cucumis Melo*). *Cucurbitaceæ.*

Seeds, sown where the plants are to stand. In the north they are occasionally started under glass in pots or pieces of inverted sods, by amateurs.

Menispermum (Moon-seed). *Menispermaceæ.*

Propagated by seeds, division or cuttings, in spring.

Mentzelia. *Loasaceæ.*

Increased by seeds in spring, in gentle heat. The seedlings of *Bartonia aurea* (properly *Mentzelia Lindleyi*) should be potted singly into small, well-drained pots. In winter they should be placed on a dry shelf in a greenhouse or frame.

Menyanthes (Buckbean). *Gentianaceæ.*

Increased by seeds; by division of the roots.

Mertensia (Lungwort). *Borraginaceæ.*

Propagation is effected by sowing seeds as soon as ripe, or by divisions in autumn.

Mesembryanthemum (Fig Marigold, Ice Plant). *Ficoideæ.*

May be easily propagated by seeds, sown under glass; by pieces, pulled or cut off and laid in the sun on moist sand.

Mespilus. See Medlar.

Michaelmas Daisy. See Aster.

Michelia. *Magnoliaceæ.*

Seeds. Make cuttings of growing wood in summer, and place in sand under glass.

Mignonette. See Reseda.

Milfoil. See Achillea.

Milkweed. See Asclepias.

Milla. *Liliaceæ.*

Increased by seeds, bulbels or by division.

Miltonia. *Orchidaceæ.*

Dividing the pseudo-bulbs. See also under Orchids.

Mimosa (Sensitive Plant). *Leguminosæ.*

Seeds, sown indoors. Cuttings of rather firm shoots, and inserted in sandy soil, in heat.

Mimulus (Monkey-flower, Musk Plant). *Scrophulariaceæ.*

Propagated by seeds, which should be thinly sown and lightly covered. Also by division, and cuttings.

Mint. See Peppermint and Spearmint.

Mirabilis, Jalapa (Marvel of Peru, Four-O'clock). *Nyctaginaceæ.*

Seeds, sown in spring either under cover or outdoors.

Miscanthus (Eulalia, Zebra-grass). *Gramineæ.*

Division and seeds.

Mistletoe. See Viscum.

Mock Orange. See Philadelphus.

Momordica. *Cucurbitaceæ.*

Increased by seeds, which should be sown in heat early in spring, or in the open in the south.

Monk's Hood. See Aconitum.

Monstera, Serangium, Tornelia. *Aroideæ.*

Easily increased by seeds and by cuttings of the stem.

Montbretia. See Tritonia.

Moon-flower. See Ipomœa.

Moon-seed. See Menispermum.

Morning-glory. See Ipomœa.

Morus. See Mulberry.

Mountain Ash. See Pyrus.

Mountain Laurel. See Kalmia.

Mourning Bride. See Scabiosa.

Mucuna (Cow-itch). *Leguminosæ.*

Propagation may be effected by seeds, or by cuttings of half-ripened wood under glass.

Muehlenbeckia, Sarcogonum. *Polygonaceæ.*

Seeds. Increased usually by cuttings, taken in early summer, in a frame.

Mulberry (*Morus alba*, *M. nigra*, *M. rubra*, etc.). *Urticaceæ.*

New sorts are grown by seeds, which should be handled in the same manner as small-fruit seeds. Named varieties are multiplied by cuttings of the root, or of mature wood, and sometimes by layers. They may be cion-budded in the spring (see Fig. 115). In the south, cuttings of the Downing mulberry are used for stocks (Fig. 105).

The common white mulberry was formerly used as a stock for named varieties, but Russian mulberry seedlings are now much used. The stocks may be top-worked outdoors (as explained above) or root-grafted in the house. The fancy varieties are commonly crown-worked, in the house in winter, the stocks being grown in pots or boxes for the purpose. They are then kept under glass until the weather permits them outdoors. By this method choice specimen trees are procured, but they are readily handled by cheaper methods. The weeping and other ornamental sorts are worked upon the Russian mulberry.

Mulberry, Paper. See Broussonetia.

Mullein. See Verbascum.

Musa (Banana, or Plantain-tree). *Scitamineæ.*

Seeds may be sown in heat during spring. Suckers are used for those species which produce them. Many of the species do not produce seeds freely, and suckers must be relied upon. *Musa Ensete* is propagated by seeds. See Banana.

Muscari, including Botryanthus (Grape Hyacinth). *Liliaceæ.*

Increased by seeds; also by bulbels, which are obtained by lifting the old bulbs early in the autumn, about every second year.

Mushroom (*Agaricus campestris*).

Break up the commercial spawn into pieces about as large as a hen's egg, and plant it two or three inches deep in drills or holes, using from one-half-pound to a pound of spawn to each square yard of bed.

The spawn is the mycelium of the fungus grown in a

mass or "brick" of earth and manure. Various methods are employed for making the spawn, but the essentials of them all are that the body of the brick shall be composed of a porous and light material, which can be compressed into a compact mass; fresh mycelium must be communicated to this mass, and then a mild heat must be applied, until the whole mass is permeated by the mycelium. The mass should be kept in heat until the whole of it assumes a somewhat cloudy look, but not until the threads of the mycelium can be seen. Ordinarily, fresh horse-manure, cow-manure and good loam are mixed together in about equal proportions, enough water being added to render the material of the consistency of mortar. It is then spread upon the floor or in large vats, until sufficiently dry to be cut into bricks. When these are tolerably well dried, mycelium from a mushroom bed or from other bricks is inserted in the side of each brick. A bit of spawn about the size of a small walnut is thus inserted, and the hole is plugged up. The bricks are now placed in a mild covered hotbed, with a bottom heat of 55° to 65°, and left there until the clouded appearance indicates that the mycelium has extended throughout the mass.

Soil from a good mushroom bed is sometimes used to sow new beds, in place of commercial spawn.

Old clumps of mushrooms may be allowed to become dry, and they may then be mixed into a bed. The spores will then stock the soil and produce a new crop. The full-grown mushroom may be laid upon white paper until the spores are discharged, and these spores may then be mixed into the earth. Propagation by spores is little understood. (See page 24.)

Mustard (*Brassica* or *Sinapis* species). *Cruciferæ.*

Propagated by seeds.

Mutisia. *Compositæ.*

Seeds. Layers and cuttings of growing wood, those of the tender species in bottom heat.

Mygindia, Rhacoma. *Celastraceæ.*

Seeds. Cuttings of firm shoots under glass.

Myosotis (Forget-me-not). *Borraginaceæ.*

Propagated by seeds sown in spring indoors or in the garden. The perennials may also be increased by division, in spring, or by cuttings placed under a hand-glass in a shady spot, in summer.

Myrica (Payberry, Sweet Gale, Wax Myrtle, Candleberry) *Myricaceæ.*

Hardy species mostly by seeds, from which the pulp has been removed. Sow as soon as ripe, or stratify them. Layers and divisions may also be employed. The greenhouse species are increased mostly by green cuttings.

Myristica (Nutmeg). *Myristicaceæ.*

May be increased by seeds; or by cuttings of ripened shoots placed in sand under glass, in bottom heat.

Myrobalan. See Prunus.

Myrrhis (Sweet Cicely or Myrrh). *Umbelliferæ.*

May be increased by divisions or by seeds.

Myrsiphyllum. *Liliaceæ.*

Freely increased by seeds, or by divisions. *M. asparagoides* (properly *Asparagus medeoloides*), the "Smilax" or Boston-vine of greenhouses, is increased by seeds, which germinate readily. The roots may also be divided, but seeds are to be preferred.

Myrtus (Myrtle). *Myrtaceæ.*

Seeds, when they can be obtained. Readily propagated by cuttings of firm or partially ripened shoots, placed in a close frame; those of the stove species require a warmer temperature than the half-hardy ones.

Nægelia. *Gesneraceæ.*

Seeds rarely. Propagation is effected by potting the runners in spring or summer in a compost of peat, leaf soil and a little loam. Cuttings of young shoots, or mature leaves, will also root readily. Compare Sinningia.

Narcissus (Daffodil, Jonquil, Chinese Sacred Lily). *Amaryllidaceæ.*

New varieties are grown from seeds, which give flowering bulbs in three or four years. Ordinarily increased by bulbels, which usually flower the second year.

Nasturtium. See Water Cress, and Tropæolum.

Nectarine. Propagated the same as Peach.

Negundo (Box Elder). *Sapindaceæ.*

Propagates with readiness by seeds, which should be sown as soon as ripe. Also by cuttings of mature wood, handled like grape cuttings.

Neillia. Handled the same as Spiræa, which see.

Nelumbo, Nelumbium (Water Chinquapin, Lotus, Water Bean). *Nymphæaceæ.*

Seeds, which may be sown in shallow pans of water in the garden, or if sown in ponds they may be incorporated in a ball of clay and dropped into the water. The seeds of some species are very hard, and germination is facilitated if they are very carefully filed or bored (see Fig. 15). Sections of the rhizomes may be used instead; they should always be covered with water, at least a foot or two deep, if outdoors. The False Lotus or Sacred Bean (*N. Indicum* or *speciosum*) by division and seeds.

Nemastylis. *Iridaceæ.*

Propagation is effected by seeds, or by bulbels.

Nemopanthes (Mountain Holly). *Ilicineæ.*

Increased by seeds, which should be sown as soon as ripe or else stratified; also by division of old plants.

Nemophila. *Hydrophyllaceæ.*

Seeds may be sown in late summer or any time during early spring.

Nepenthes (Pitcher Plant). *Nepenthaceæ.*

Propagated by seeds and cuttings. The seeds must have good drainage, uniform conditions and strong heat (80° to 85°). Sow upon a soil made of peat and fine sphagnum, and keep in a moist and close frame. Cuttings are usually struck in moss in a frame having strong bottom heat. A good plan is to fill a small pot with moss, invert it, and insert the cutting through the hole in the bottom. The pot then keeps the moss uniform in temperature and moisture. The pot is broken when the plant is removed. When potting off, use very coarse material.

Nepeta, Glechoma (Catmint, Catnip). *Labiatæ.*

Propagated by sowing seed in spring, or by division.

Nephelium (Litchi). *Sapindaceæ.*

May be increased by seeds, or by cuttings made of half-ripened wood.

Nephrodium. See Ferns.

Nephrolepis. See Ferns.

Nerine (Guernsey Lily). *Amaryllidaceæ.*

Seeds, for new varieties. Commonly increased by means of bulbels.

Nerium (Oleander). *Apocynaceæ.*

Layers. Cuttings should be made of natural leading shoots, inserted in single pots and placed in a close, warm frame; or they may be rooted in bottles of water and afterwards potted in soil. See Fig. 69 *a*.

Nertera. *Rubiaceæ.*

Increased by seeds, divisions, or cuttings. Any small portion will grow freely, especially if placed in a warm frame.

Nettle-tree. See Celtis.

New Zealand Flax. See Phormium.

Nicandra. *Solanaceæ.*

Seeds, sown in the open border, or under glass in the north.

Nicotiana (Tobacco). *Solanaceæ.*

Propagated by seeds, started under glass or in a carefully prepared seed-bed. The ornamental species sometimes by cuttings. *N. alata* (*N. affinis* of gardens) propagates by root cuttings. Tobacco is handled essentially like tomato plants.

Nierembergia. *Solanaceæ.*

Grown from seeds, under glass. Cuttings of firm shoots are also used.

Nigella (Fennel Flower, Love-in-a-mist). *Ranunculaceæ.*

Propagated by seeds sown in early spring in the open.

Nightshade. See Solanum.

Nine-bark. See Spiræa.

Nolana. *Convolvulaceæ.*

Seeds sown in the open border during spring.

Norfolk Island Pine. See Araucaria.

Norway Spruce. See Picea.

Nuphar. Propagated same as Nelumbo and Nymphæa, which see.

Nutmeg. See Myristica.

Nuttallia. *Rosaceæ.*

May be propagated by seeds; by divisions; by means of suckers, which spring from the roots.

Nut-trees. See the various genera, as Almond, Chestnut, Hicoria, Juglans and Pecan.

Nymphæa, Castalia (Water Lily, Lotus). *Nymphæaceæ.*

Seeds, which are rolled up in a ball of clay and dropped into a pond, or sown in pots which are then submerged in shallow water, either indoors or out. Usually increased by portions of the rootstocks, which are sunk in the pond and held by stones, or the tender species placed inside, in pans of water. Some species produce tubers on the rootstocks, which are used for propagation.

Nyssa (Pepperidge, Sour Gum, Tupelo-tree). *Cornaceæ.*

Increased by seeds and by layers. The seeds should be sown as soon as ripe or else stratified. They usually lie dormant the first year.

Oak. See Quercus.

Obeliscaria. See Rudbeckia.

Ochna. *Ochnaceæ.*

During summer, cuttings may be made of growing shoots.

Odontoglossum. *Orchidaceæ.*

Division. See also under Orchids.

Œnothera, including Godetia (Evening Primrose). *Onagraceæ.*

Seeds may be sown in spring or summer. Divisions may be made. Cuttings of perennials should be placed in a cool frame in the early part of the season before flowering begins.

Okra, Gumbo (*Hibiscus esculentus*). *Malvaceæ.*

Seeds, sown where the plants are to stand, or started in pots often in the north.

Olax. *Olacineæ.*

Grown from cuttings of firm or mature shoots in heat.

Olea. *Oleaceæ.*

The ornamental species are grown from cuttings of ripened shoots, either under frames or in the border, and also by seeds. For propagation of *O. Europæa*, see Olive.

Oleander. See Nerium.

Oleaster. See Elæagnus.

Olive (*Olea Europæa*). *Oleaceæ.*

The olive is grown in large quantities from seed, especially in Europe. The pulp is removed by maceration or by treating with potash. The pits should be cracked or else softened by soaking in strong lye, otherwise they will lie dormant for one or two years. Cuttings of any kind will grow. Limbs, either young or old, an inch or two inches in diameter, and from 1 to 2 feet long, are often stuck into the ground where the trees are to grow, or they are sometimes used in the nursery. Green cuttings, with the leaves on, are often used, being handled in frames or in boxes of sand. Chips from old trunks, if kept warm and moist, will grow. The olive is often propagated by truncheons of trunks. A trunk 2 or 3 inches in diameter is cut into foot or two-feet lengths, and each length is split through the middle. Each half is planted horizontally, bark up, 4 or 5 inches deep, in warm moist soil. The sprouts which arise may be allowed to grow, or they may be made into green cuttings. Knaurs (see page 64) are sometimes used. The olive can be budded or grafted in a variety of ways. Twig-budding and plate or H-budding (Figs. 96, 97, 98) give admirable results, and are probably the best methods. Twig-budding is the insertion of a small growing twig which is cut from the branch in just the manner in which shield-buds are cut. (Fig. 96.) Side-grafting is also successful. (Fig. 113.)

Omphalodes, Picotia. *Borraginaceæ.*

Freely increased by means of seeds planted in spring, or by division.

Oncidium. *Orchidaceæ.*

Division. In some species detachable buds are produced in the inflorescence, and these give young plants. (See also under Orchids.)

Onion (*Allium Cepa* and *A. fistulosum*). *Liliaceæ.*

Onions are mostly grown from seeds, which must be

sown as early as possible in spring; or in the south they may be sown in the fall. They are also grown from "tops," which are bulblets borne in the flower cluster. These are planted in the spring, or in the fall in mild climates, and they soon grow into large bulbs. "Sets" are also used. These are very small onions, and when planted they simply complete their growth into large bulbs. Sets are procured by sowing seeds very thickly in poor soil. The bulbs soon crowd each other, and growth is checked, causing them to ripen prematurely. Good sets should not be more than a-half inch in diameter. Very small onions which are selected from the general crop—called "rare-ripes"—are sometimes used as sets, but they are usually too large to give good results. Some onions—the "multiplier" or "potato onions"—increase themselves by division of the bulb. The small bulb, which is planted in the spring, splits up into several distinct portions, each one of which will multiply itself in the same manner when planted the following year.

Onobrychis (Saintfoin). *Leguminosæ.*

Seeds, sown in spring where the plants are to remain.

Onosma (Golden Drop). *Borraginaceæ.*

Seeds, sown in the open in spring. Perennial species by cuttings in summer.

Opuntia (Prickly Pear, Indian Fig). *Cactaceæ.*

Seeds grow readily, sown as soon as ripe in ordinary sandy soil, either in the house or outdoors. The joints grow readily if laid on sand. It is customary to allow these cuttings to dry several days before planting them. See also Cactus.

Orach (*Atriplex hortensis*). *Chenopodiaceæ.*

Seeds, sown where the plants are to stand.

Orange (*Citrus Aurantium*, etc.). *Rutaceæ.*

Orange stocks are grown from seeds, which should be cleaned and stratified in sand or other material, until sowing time. The seeds should not be allowed to become hard and dry. Some prefer to let the seeds sprout in the sand and then sow them in the nursery, but they must be carefully handled. The seeds are usually sown in seed beds, after the manner of apple seeds, and the seedlings are transplanted the next fall or spring into nursery rows. Care must always be exercised in handling orange plants

Orange, continued.

as they are often impatient of transplanting. Oranges grow readily from cuttings, although cuttage is not often practiced. Green cuttings, handled under a frame, give good results. Mature wood, either one or two years old, can be treated after the manner of long grape cuttings. They must have an abundance of moisture. Layers are sometimes made.

The named varieties are shield-budded upon other stocks. Grafting can be practiced, but it is often unsatisfactory. The nursery stocks are commonly budded in the spring, after having grown in the rows one year, which is two years from the sowing of the seed. If thorn-bearing varieties are to be propagated, a thorn with a bud in its axil is often cut with the bud, to serve as a handle in place of the leaf-stalk, which is used in summer budding. Many stocks are used for the orange. The leading ones are sweet or common orange, sour orange (*Citrus Aurantium*, var. *Bigaradia*), pomelo (*Citrus decumana*), Otaheite orange, trifoliate orange (*Citrus*, or *Ægle*, *trifoliata*), and various lemons, as the "French" or Florida Rough and the Chinese. For general purposes, the sweet and sour orange stocks are probably the best. The sour stock is obtained from wild seeds, this variety having extensively run wild in Florida from early times. The trifoliate and Otaheite stocks are used for dwarfing or for small growing sorts, as many of the Japanese varieties. The trifoliate orange is also one of the hardiest of the orange stocks, and its use will probably increase upon the northern limit of the orange belt. Old orange trees can be top-budded with ease. It is advisable to cut them back a year before the operation is performed, in order to secure young shoots in which to bud. In ordinary greenhouse practice, the seedlings of the pomelo make good stocks. They can be established in three-inch pots the first season, and veneer-grafted the next winter.

The Rowell method of propagating the orange (so named for William M. Rowell, Fort Meade, Florida, its inventor) is thus described by a local Florida newspaper (Bartow Courier-Informant, 1891):

"Mr. Rowell's process is almost startling in novelty, yet it is very simple. Briefly stated, it is about as follows: Cuttings ¼ to ½ inch in diameter and 10 or 12 inches long, are taken from any healthy citrus tree, and buds of any desired variety are put in them. This is done in the house or barn, and as the cuttings are budded they are

Orange, concluded.

placed in boxes and lightly covered with dirt. There they remain until wanted for planting. The cuttings will form roots, but the buds will remain dormant until the cuttings are transplanted, whether that be three weeks or three years.

"When planting in grove form, the cuttings are placed in a vertical position if seedling trees are to be imitated, or in an almost horizontal position if it be the grower's intention to plant close and produce small trees; and when the object is to dwarf the trees, the cuttings are almost inverted. In either case, the cutting is entirely covered with dirt, except the portion occupied by the bud, which is protected by a small cylinder of zinc, 2 or 3 inches long, which is fitted to the cutting and protrudes through the soil, giving light and air to the bud. This is removed, however, when the bud attains a height of 10 or 12 inches, and the soil is then drawn up around the bud. The subsequent cultivation is the same as with trees propagated in the usual way. Mr. Rowell has applied for a patent on the tube.

"Now for results. Mr. Rowell has a grove which has been produced by his method. It is on new pine land that has never been cowpenned or fertilized in any way. The grove is now yielding its first crop—over 300 oranges to the tree in some instances—and is only three years old. The public is invited to inspect these trees and compare them with any well cared-for seeedlings six years old. There are some other advantages claimed for this system of propagation which we cannot now point out.

"Mr. Rowell has Japan persimmons budded in the same way."

Orchids. *Orchidaceæ.*

The method of propagating these plants must in each species be adapted to the habit and mode of growth. The easiest and safest plan for the vast majority is by division, but seeds, cuttings, layers, offsets, and very rarely roots, are also utilized. It is important that artificial means of increase should only be adopted where the individual plants are in robust health. With many orchids the struggle of life under the unnatural conditions we supply, is necessarily severe, and any operation which transforms one weak plant into two or more weaker ones, is to be deprecated. In cases where the only method available necessitates disturbance at the roots, consideration must

Orchids, continued.

be paid to the constitution of the species, for some orchids, even when perfectly healthy, strongly resent interference.

Seeds.—In no class of cultivated plants is propagation by seeds more difficult and tedious than it is with orchids. In all cases, fertilization must be performed by hand. In England, the length of time required for the capsules to ripen varies from three months to a year. Good seeds form a very small proportion of the whole, and it occasionally happens that the contents of a capsule will not produce a single plant. This, however, as well as the difficulty experienced in England in rearing plants to the flowering stage, is primarily due to the deficiency of sunlight, and in such a bright climate as that of the United States, would not be likely to occur. Various methods of sowing are in vogue, such as sprinkling over pieces of wood and cork or tree-fern stem, and on the top of moss and peat, in which established plants of the same or a nearly related species are growing. The last is probably the best, but it is always advisable to try several methods. Of course, the material on which the seeds are scattered must always be kept moist and shaded. The period between germination and the development of the first root is the most critical in the life of a seedling orchid. After they are of sufficient size to handle they are potted off into tiny pots, and as they gain strength, are given treatment approximating that of adult plants.

Division.—Cypripediums may be taken as an example where this is readily done. It is simply necessary to carefully shake off the soil from the roots, and by the aid of a sharp knife, sever the plant into as many pieces as are required. It is always advisable to leave one or more leading growths to each portion. This method may be practiced for the increase of phaius, masdevallia, sobralia, ada, the evergreen section of calanthe, and all of similar habit. In nearly all those kinds where the pseudo-bulbs are united by a procumbent rhizome, such as occurs in cattleyas, the process is slower. It seems to be natural for these plants to continue year after year, producing a single growth from the old pseudo-bulb. To obtain additional "leads," the rhizomes should be cut through in early spring, two or three pseudo-bulbs being reserved to each piece. A bud will then push from the base of each pseudo-bulb nearest the division, and a new lead is formed. The pieces should not be separated until this is well established, and three years may sometimes be re-

Orchids, concluded.

quired. Lælia, catasetum, cœlogyne, lycaste, cymbidium, zygopetalum, odontoglossum, oncidium, miltonia, etc., are treated in this manner.

Cuttings.—This method is available for those kinds with long, jointed stems, like dendrobium and epidendrum. Just before the plants commence to grow, say in February, the old pseudo-bulbs are cut up into lengths, and laid on a moist, warm surface, such as a pan of moss in a propagating frame. Young offshoots will shortly appear at the nodes, and when large enough are potted off with the old piece attached. This plan may be used also for barkeria and microstylis.

It is well to remember that in any method of propagation where the pseudo-bulb is divided, the vigor of the young plant is proportionate to the amount of reserve material supplied it. However suitable the external conditions may be for growth, it is for some time entirely dependent for sustenance on the old piece from which it springs. *Dendrobium Phalænopsis* is a case in point. If a pseudo-bulb is cut into say three pieces, it will take at least two years for the young plants to reach flowering strength, but frequently, by using the entire pseudo-bulb, we can get in a single year a growth quite as large as the old one.

The treatment of young orchids should be founded on what suits the parents. As a rule, however, they require more careful nursing, and some of the conditions must be modified. Drought, intense light and cold draughts must be avoided. For many orchids, especially those from equatorial regions, where the atmospheric conditions alternate between saturation and intense heat and dryness, it is necessary, in order to induce flowering, that nature, to some extent at least, should be imitated. With young plants, by whatever method they may be obtained, the supply of water must only be reduced in accordance with the weather and season, and beyond that, no attempt at resting made. In cases, however, where plants have been divided or made into cuttings, a very limited supply of water is needed at first; but to prevent exhaustion, the atmosphere should always be kept laden with moisture.

Oreopanax. *Araliaceæ.*

Seeds, and cuttings of the young shoots, or division of well-established plants.

Ornithogalum (Star of Bethlehem). *Liliacæ.*

Seeds. Commonly by bulbels, and by division.

Orobus (Bitter Vetch). *Leguminosæ.*

Readily propagated by seeds, or by dividing the tufts.

Orontium. *Aroideæ.*

Commonly increased by division, but seeds may be used.

Orpine. See Sedum.

Osage Orange. See Maclura.

Osier. See Salix and Cornus.

Osmanthus (Japan Holly). *Oleaceæ.*

Propagated by cuttings under glass, or by grafting on osmanthus stock, or on privet.

Osmunda (Flowering Fern). *Filices.*

Mostly by division; sometimes by spores. See Ferns.

Ostrowskia (Giant Bellwort). *Campanulaceæ.*

Propagated in the same manner as the perennial campanulas, which see.

Ostrya (Hop Hornbeam, Ironwood). *Cupuliferæ.*

Best grown from seeds. Also increased by layering; or it can be grafted. The European species is often grafted upon the hornbeam (carpinus).

Othonopsis, Othonna (Ragwort). *Compositæ.*

Very easily propagated by seeds and cuttings. The leaves also take root.

Ouvirandra (Lattice-leaf). *Naiadaceæ.*

The plants are divided; or seeds are used when they can be obtained.

Oxalis. *Geraniaceæ.*

Seeds, divisions and cuttings. The tuberiferous species are increased by the small tubers upon the roots.

Oxydendrum (Sorrel-tree). *Eriacaceæ.*

Increased by seeds, which must be handled carefully in light soil. Also by layers, which, however, often root with difficulty.

Oxylobium (Callistachys). *Leguminosæ.*

Cuttings of firm wood, in spring, under glass.

Oxytropis. *Leguminosæ.*

Seeds should be sown where the plants are to stand; also by dividing the plant in spring.

Oyster Plant. See Salsify.

Pæony (Peony, Piney). *Ranunculaceæ.*

Seeds, giving new varieties, are sown as soon as ripe. The seedlings seldom rise above the surface the first year, all their energies being spent in the formation of roots. The common herbaceous varieties are oftenest propagated by division of the clumps. Each portion should possess at least one bud upon the crown. All woody species may be increased by layers and cuttings. Cuttings are taken late in summer, cut to a heel, and are handled in a frame or cool greenhouse. During winter they should be kept from freezing. The shrubby species and *P. Moutan* are often grafted, and all species can be handled in this way. The operation is performed in late summer or early autumn, and the grafts are stored in sand or moss where they will not freeze. The next spring they are planted out. The cion is made from a strong short shoot, destitute of flower buds, and is set upon a piece of root, as described on pages 128, 129. Some prefer to cut a wedge-shaped portion from the side of the stock, in which to inlay the cion, rather than to split the stock; but either practice is good. Strong roots of various varieties or species may be used. The Chinese pæony (*P. Moutan*), *P. officinalis* and *P. albiflora* are oftenest used.

Paliurus, Aubletia (Christ's Thorn). *Rhamnaceæ.*

May be increased by seeds, by layers or by cuttings of the roots.

Palma-Christi. See Ricinus.

Palmetto. See Sabal, and Palms.

Palms. *Palmaceæ.*

Palms are mostly grown from imported seeds. These should always be sown in a brisk bottom heat, in a mixture of coarse loam and sand. A hotbed, established upon the greenhouse bench, is an excellent place in which to start palm seeds. Some species are increased by suckers, which arise from the crown or roots. For more explicit directions, see the various genera.

Pampas Grass. See Gynerium.

Panax. See Ginseng.

Pancratium and **Hymenocallis.** *Amaryllidaceæ.*

Seeds, sown in pans in heat, are sometimes employed. Commonly increased by offsets, which usually form freely.

Pandanus (Screw Pine). *Pandanaceæ.*

Seeds and suckers, as in palms. Also by cuttings of the young growth in heat. The "seeds" are really fruits, and if in good condition several plants, one to ten, are obtainable from each; they should be separated when well furnished with roots. These seeds are easily obtained in the tropics, and are planted in moist black soil in beds or pots. When the plants appear, the little clumps are separated and the plantlets potted off.

Pansy. See Viola.

Papaver (Poppy). *Papaveraceæ.*

Seeds—usually sown outdoors—and divisions. *P. orientale* and allied species are easily propagated by root-cuttings in sandy soil under glass in autumn.

Papaw-tree. See Carica; also Asimina.

Papyrus. *Cyperaceæ.*

Propagation by seeds and by divisions, chiefly the latter.

Pardanthus. See Belamcanda.

Paris. *Liliaceæ.*

Increased by seeds or by divisions.

Paris Daisy. See Marguerite.

Parkinsonia. *Leguminosæ.*

Seeds mostly. Cuttings.

Parnassia (Grass of Parnassus). *Saxifragaceæ.*

May be propagated by seeds or by divisions.

Parrotia. *Hamamelideæ.*

Increased by seeds or by layers.

Parsley (*Apium Petroselinum*). *Umbelliferæ.*

Seeds, which are usually sown outdoors. The roots may be taken up in fall to be forced under glass.

Parsnip (*Pastinaca sativa*). *Umbelliferæ.*

Fresh seeds, sown where the plants are to stand.

Pasque-flower. See Anemone.

Passiflora (Passion Flower). *Passifloraceæ.*

Seeds, sown under glass. Cuttings of the young growth root easily in sand in a frame. Varieties are sometimes veneer-grafted, *e. g.*, *P. coccinea. P. cœrulea* propagates by root-cuttings.

Paulownia. *Scrophulariaceæ.*

Seeds, sown in carefully prepared soil, either in a seed bed or in a coldframe. Cuttings of ripe wood or of roots made in fall or spring.

Pea (*Pisum sativum*). *Leguminosæ.*

Seeds, sown where the plants are to stand. The plants are hardy and seeds may be sown very early. For Cow-Pea, see Vigna.

Peach (*Prunus Persica*). *Rosaceæ.*

The peach is perhaps the easiest to propagate of all northern fruit trees. Stocks are universally grown from seeds, although root-cuttings will grow. The seeds should be buried outdoors in the summer or fall, and shallow enough so that they will be fully exposed to frost. Some prefer to simply spread them upon the surface of the ground and cover them lightly with straw to prevent them from drying out. The pits should be kept moist, and by spring most of them will be cracked. Those which do not open should be cracked by hand, for if planted they will not germinate until a year later than the others. In large nurseries, however, the cracking of peach pits by hand is too expensive to be practiced. The "meats" or kernels are sorted out and planted early in drills. Some prefer to sprout the seeds in the house, in order to select the best for planting. Some growers upon a small scale pinch off the tip of the rootlet to make the root branch. Pits should be secured, of course, from strong and healthy trees, but the opinion that "natural seed," or that from unbudded trees, is necessarily best, is unfounded.

The seeds should be planted in rich soil, and the stocks will be large enough to bud the same year. Any which are not large enough to bud may be cut back to the ground the next spring, and one shoot be allowed to grow for budding, but such small stocks are usually destroyed, as it does not pay to bestow the extra labor and use of land upon them. When the buds have grown one season, the trees are ready for sale—at one year from the bud and two years from the seed. Peach trees should never be more than a year old (from the bud) for orchard planting. June-budded trees are much used in the south (see page 103). Peach trees are always shield-budded, and the operation is fully described on pages 95 to 105. Grafting can be done, but as budding is so easily performed, there is no occasion for it. The peach shoots are so pithy that, in making cions, it is well to leave a portion of the old wood upon the lower end—extending part way up the cut—to give the cion strength. Peach wounds heal so slowly and imperfectly that grafting is never to be recommended.

Peaches are nearly always worked upon peaches in this country. Plums are occasionally employed for damp and strong soils. Myrobalan plum is sometimes used, but it cannot be recommended. All plums dwarf the peach more or less. The hard-shell almond is a good stock for very light and dry soils. The Peen-to and similar peaches are worked upon common peach stocks.

The ornamental peaches are budded upon common peach stocks in the same manner as the fruit-bearing sorts.

The nectarine is propagated in exactly the same manner as the peach.

For *Prunus Simoni*, see Plum.

Pea-nut. See Arachis.

Pear, Alligator or **Avocado.** See Persea.

Pear (*Pyrus communis, P. Sinensis*). *Rosaceæ.*

Pear seedlings are grown in the same manner as those of the apple, which see. Pear stocks are imported from France, however, as the leaf-blight is so destructive to them here as to render their culture unprofitable. This leaf-blight is a fungus (*Entomosporium maculatum*), and recent experiment has shown that it can be readily overcome by four or five thorough sprayings with Bordeaux mixture, so that there is reason to hope that the growing of pear stocks may yet become profitable in this country, although the higher price of labor here, and the drier summers, are serious disadvantages. Heretofore, the only means of mitigating the ravages of this blight was the uncertain one of inducing a strong growth early in the season. Even when pear stocks are raised in this country, they are grown from imported French seed. Aside

Pear, continued.

from its cheapness, however, this foreign seed probably possesses no superiority over domestic seed. But pear seed is so difficult to obtain in America that it is practically out of the market. Seedlings of the sand pear type have been strongly recommended for stocks, but they do not attain general favor amongst nurserymen.

Pear seedlings should be taken up and removed from the seed-bed the first fall. The foreign stocks are imported when a year old from the seed. The seedlings are trimmed and sometimes "dressed" (see page 96), and are set into nursery rows the following spring. The next season — that is, the season in which the stocks are transplanted — shield-budding is performed, as upon the apple. The budding season usually begins late in July or early in August in the north. If the stocks are small, of "second size," they may stand over winter and be budded the second year. Pear trees are sold at two and three years from the bud. Pears do not succeed well when root-grafted, except when a long cion is used, for the purpose of securing own-rooted trees (see page 110). Dormant buds of the pear may be used upon large stocks in early spring, the same as upon the apple, and buds may be kept upon ice for use in early summer (see page 103).

Pears are dwarfed by working them upon the quince. The Angers quince is the best stock. The ordinary orange quince and its kin generally make weak and short-lived trees. Quince stocks are obtained from ordinary cuttings or from mound-layering, the latter method giving much the better stocks (see Quince). The layers should be removed the first autumn; or, if they are not rooted then, they may be left a year longer, when they will be found to be well rooted, and may then be taken off, trimmed up and fitted to plant as stocks the following spring, and budded in August. Quince stocks are bought in Europe, whence they arrive in the fall. They are "dressed" and set in nursery rows the following spring, and the buds are set during the first season. It is imperative to set the bud as low as possible in order to secure trees which can easily be set deep enough to cover the union (4 to 6 inches below the surface is the common depth of planting dwarf pears). Some varieties do not unite well with the quince, and if it is desired to dwarf them, they should be double-worked (see page 133). Some of the common and popular varieties which thrive directly upon the quince (without double-working) are the

Pear, concluded.

following: Angouleme (Duchess), Anjou, Louise Bonne, Howell, White Doyenne (Virgalieu), Manning's Elizabeth, Lawrence. Varieties which usually thrive better when double-worked are Clairgeau, Bartlett, Seckel, and others.

The pear can also be grown upon the apple, thorn and mountain ash. Upon the apple it is short-lived, although pear cions, set in the top of an old apple tree, often bear large fruits for a few years. When pear stocks cannot be had, pears are sometimes worked upon apple roots. If the cions are long they will emit roots, and when the apple nurse fails the pear becomes own-rooted. Good dwarf trees are often secured upon the thorn, and there is reason to believe that some of the thorns will be found to be preferable to quince stocks for severe climates and for special purposes. The subject is little understood. The mountain ash is sometimes used for the purpose of growing pears upon a sandy soil, but its use appears to be of little consequence.

Pears of the Le Conte and Kieffer type are often grown from cuttings in the south. Cuttings are made of the recent mature growth, about a foot in length, and are planted in the open ground, after the manner of long grape cuttings. Le Conte, Garber, Smith, and other very strong growers of the Chinese type, are probably best when grown from cuttings. They soon overgrow French stocks, as also apple stocks, which have been used to some extent; but if long cions are used, own-rooted trees are soon obtained, and the stock will have served a useful purpose in pushing the cion the first two or three years.

Pecan (*Hicoria Pecan*). *Juglandaceæ*.

Propagated by seeds. These may be planted as soon as ripe, or stratified until spring. The ground should be well prepared, and the nuts planted about 3 inches deep. By grafting on pecan or common hickory stock that is not over 2 years old. Cions about 6 inches long should be cut during the winter and put in a cool place to hold them back until the stocks have fairly started in the spring. The stalks should then be cut off at (or preferably 3 to 6 inches below) the crown, and the cion inserted. The tongue-graft gives the best result, although, as in all nuts, the skill of the grafter is more important than the method. Bandage securely, and bank with earth nearly to the top of the cion, to keep it moist. It can also be budded, like peaches. The pecan and other hickories will also grow from cuttings of the ends of the soft growing roots. See Hicoria.

Pelargonium (Geranium, Stork's Bill). *Geraniaceæ.*

Seeds, sown in light soil with mild heat, are sometimes employed. Commonly increased by cuttings of firm shoots, which grow readily (Figs. 69 *c*, and 73). The common geraniums, for conservatory use, should be renewed from cuttings every year. The fancy or show geraniums are often grown from root-cuttings, but sometimes will not come true. Geraniums can also be grafted. (See page 130, herbaceous grafting.)

Pelecyphora (Hatchet Cactus). *Cactaceæ.*

Propagated most freely by seeds in moderate heat, and by cuttings made of any small shoots that arise from the base. See also Cactus.

Peltandra. *Aroideæ.*

Propagated by seeds when fresh, or by division.

Pennyroyal (*Mentha Pulegium*). *Labiatæ.*

Seeds and division.

Pentstemon (Beard-tongue). *Scrophulariaceæ.*

Seeds, sown in pans and placed under a frame; or they are sometimes sown in the border where the plants are to stand. Also by division, and rarely by cuttings in summer.

Peony. See Pæonia.

Peperomia, including Micropiper. *Piperaceæ.*

Seeds. Cuttings of single joints of firm stems root easily in a peaty soil. Water sparingly.

Pepper, Black. See Piper.

Pepperidge. See Nyssa.

Pepper-grass, Curled Cress (*Lepidium sativum*). *Cruciferæ.*

Grown from seeds, either under glass for early crops or in the open air.

Peppermint (*Mentha piperita*). *Labiatæ.*

Divisions of the creeping and rooting stems are planted to multiply the plant, and plantations are renewed every three or four years.

Pepper, Red or **Cayenne** (*Capsicum*). *Solanaceæ.*

Seeds, sown outdoors, or in the north oftener started in the house.

Pereskia (Barbadoes Gooseberry). *Cactaceæ.*

Seeds. Cuttings, as described under Cactus. *P. aculeata* is much used as a stock for epiphyllums. *P. Bleo* is sometimes used for the same purpose, as it is fully as good as the other species. Cuttings of *P. aculeata* can be made a foot or more in length, and of sufficient size for immediate use; or, the graft may be inserted when the cutting is made.

Perilla. *Labiatæ.*

Sow the seeds in early spring in pans or boxes, and place in a gentle heat. Or southwards, seeds may be sown in the open.

Periploca. *Asclepiadaceæ.*

Seeds. Increased mostly by layers or cuttings under glass, during summer or autumn. Root cuttings succeed.

Periwinkle. See Vinca.

Persea (Alligator or Avocado Pear). *Lauraceæ.*

Seeds. Layers of ripened shoots may be made in autumn; or cuttings of firm shoots in spring, under glass.

Persimmon (*Diospyros Kaki* and *D. Virginiana.*) *Ebenaceæ.*

Stocks are readily grown from seed, and they usually attain sufficient size for budding the first year. The native persimmon (*Diospyrus Virginiana*) is largely used as a stock for the Japanese persimmon or kaki. Imported stocks are occasionally employed, but the native is more vigorous, as a rule, and probably better. Persimmons are shield-budded the same as peaches, and they may be root-grafted and top-grafted by ordinary methods. The Rowell method of propagating Japanese persimmons is described under Orange.

Persoonia (Linkia). *Proteaceæ.*

Propagated by cuttings of the ripened shoots under glass.

Peruvian Bark. See Cinchona.

Petalostemon (Prairie Clover). *Leguminosæ.*

Seeds and divisions.

Petunia. *Solanaceæ.*

Seeds, either indoors or in the garden. Choice and double varieties are often increased by cuttings, which grow readily.

Phacelia, Eutoca, Whitlavia. *Hydrophyllaceæ.*

The annuals are increased by seeds, and the perennials by seeds and division.

Phaius. *Orchidaceæ.*

Division of the bulbs. See also under Orchids.

Phalænopsis. *Orchidaceæ.*

These are very slow and difficult to propagate. In the majority of the species it can only be done where a lateral offshoot is made from the main stem. Some species, such as *P. Luddemanniana*, and more rarely *P. amabilis*, *P. Stuartiana* and *P. Schilleriana*, develop plantlets on the old flower scapes. By pegging these down on a basket of moss they may be established and afterwards separated. *P. Stuartiana* and *P. deliciosa* have been known to produce plants on the roots. Other instances of root-proliferation are recorded in *Saccolabium micranthum* and a species of cyrtopodium. See under Orchids.

Phalaris. *Gramineæ.*

Propagated by seeds, but the sports or varieties by division.

Phaseolus (Bean, Kidney, Pole, String, Lima, French Bean, etc.). *Leguminosæ.*

The ornamental greenhouse kinds are grown from seeds planted in light soil in a warm propagating house. See Bean.

Phellodendron (Cork Tree). *Rutaceæ.*

Increased by seeds, layers, and by root cuttings.

Philadelphus (Mock Orange, Syringa). *Saxifragaceæ.*

Seeds, layers, suckers, and cuttings. Layers are sometimes used. Cuttings of mature wood are generally employed. Some well-marked varieties, like vars. *nana* and *aurea* of *P. coronarius*, are grown from cuttings of soft wood in summer in frames.

Phillyrea (Jasmine Box, Mock Privet). *Oleaceæ.*

Seeds. May be propagated by cuttings, layers, or by grafting on the privet.

Philodendron. *Aroideæ.*

Increased by seeds; and by dividing the stems, allowing two or three joints to each piece, inserting them in pots in a brisk heat.

Phlomis. *Labiatæ.*

All of the species may be increased by seeds; the herbaceous kinds by division, and the shrubby sorts also by cuttings.

Phlox. *Polemoniaceæ.*

The annuals are grown from seeds sown in the open. The perennials are grown from seeds, divisions, cuttings of stems and roots. Cuttings made during summer, and handled in a frame, do well. The roots are cut into short pieces, and are then handled in pans or flats under cover.

Phœnix. See Date.

Phormium (Flax Lily, or New Zealand Flax). *Liliaceæ.*

Seeds. Also by division of the crowns before growth commences in spring.

Photinia, including Eriobotrya. *Rosaceæ.*

Stratified seeds or half-ripened cuttings under glass. Varieties of loquat, *P. Japonica*, are grown from layers or cuttings of ripe wood; it is also worked upon seedling stocks or upon thorn or quince, after the manner of pears.

Phyllanthus, including Xylophylla. *Euphorbiaceæ.*

Increased by means of cuttings of hard shoots in heat.

Phyllocactus, including Phyllocereus and Disocactus (Leaf Cactus). *Cactaceæ.*

Seeds germinate readily in sandy soil. Usually increased by cuttings of the stems, 5 or 6 inches long, placed in sandy soil, which is kept only slightly moist. See also Cactus.

Phyllocladus. *Coniferæ.*

Cuttings of the ripened shoots under glass, in spring. When the cuttings begin callusing, give mild bottom heat.

Physalis (Ground or Winter Cherry, Strawberry Tomato, Husk Tomato). *Solanaceæ.*

Seeds, sown outdoors or under cover. Perennials by division and soft cuttings.

Phyteuma, Rapunculus (Horned Rampion). *Campanulaceæ.*
Easily increased by seeds or by division, in spring.

Phytolacca (Spoke, Skoke, Poke). *Phytolaccaceæ.*

May be propagated by means of seeds, or by division.

Picea (Spruce). *Coniferæ.*

Propagated by seeds, sometimes by layers, or grafts. Seedlings must be shaded the first year. Also by cuttings of recent wood (Fig. 67 and page 64). The spruces are easily grafted. *P. excelsa* (Norway spruce) makes a good stock; the veneer-graft, under glass, in winter, succeeds better than any method of outdoor work practicable in our climate; if the graft is inserted near the base in young plants, it is quite possible to obtain them on their own roots after a few transplantings. Side shoots can be used as cions, and if started in time will furnish good leaders; sometimes a leader is developed more rapidly by bending the plant over at nearly a right angle, when a stout bud may start from the stem. The Balsam fir is also a good stock. See Abies.

Pickerel Weed. See Pontederia.

Picotee. See Dianthus and Carnation.

Pie-plant. See Rheum.

Pilea (Artillery Plant, Stingless Nettle). *Urticaceæ.*

May be increased by seeds, division or cuttings, commonly the last.

Pilocereus. See Cactus.

Pimpernel. See Anagallis.

Pinanga. *Palmaceæ.*

Propagated by seeds.

Pinckneya. *Rubiaceæ.*

Seeds. Cuttings of the ripened shoots under glass.

Pine-apple (*Ananas sativus*). *Bromeliaceæ.*

Pine-apples very rarely produce seeds, but when they are produced they are sown for the purpose of obtaining new varieties. The pine-apple is usually increased by suckers and "crowns." If the root is left in the ground after the pine is removed, suckers will start from it. The root is then taken up and cut into as many pieces as there are suckers, each piece being then permanently planted.

The crown of the fruit and the various offsets or "crownlets," which appear on the sides and base of the fruit, may be removed and used as cuttings. These offsets are commonly used in greenhouse propagation. It is the usual practice to allow them to dry several days before they are planted, and in pine-apple regions they are often exposed to the sun for several weeks. This operation is unnecessary, however, although it is not objectionable. A good way to start the offsets is to pull off the lowest leaves and insert the offsets in damp moss in shade—giving bottom heat for greenhouse work—and as soon as roots begin to form, which will occur in from two to six weeks, plant them out permanently. In the tropics fruit can be obtained in twenty months after the offsets are transplanted; but fruit bearing is often delayed three or four years under poor treatment.

Piney. See Pæonia.

Pinguicula (Butterwort). *Lentibulariaceæ.*

The hardy and greenhouse species are increased by seeds, division, or by leaf cuttings.

Pink (*Dianthus*, various species). *Caryophyllaceæ.*

Seeds and division. Best results by raising plants from seed every two or three years. Seeds are usually sown where the plants are to remain; or they may be sown in a coldframe and transplanted. See Carnation and Dianthus.

Pinus (Pine). *Coniferæ.*

Seeds, which should be kept dry over winter, are commonly employed. These are often started in pots, but for most species they are sown in well prepared beds outdoors. The seedlings must usually be shaded the first season. Varieties, as also species which do not produce seed freely, may be grafted upon stocks of white or Austrian pine or other species. This grafting may be done upon the tips of growing shoots early in the season (page 131), but it is oftener performed upon potted plants by the veneer method.

Piper, Cubeba (Pepper, Cubeb). *Piperaceæ.*

Seeds. All are increased by means of cuttings of the growing shoots, inserted in sandy soil under glass.

Piqueria. *Compositæ.*

Piqueria trinervia (the *Stevia serrata* of florists) is generally grown from cuttings, like fuchsias and carnations. Also by seeds, which are freely produced.

Pistacia. *Anacardiaceæ*

Seeds, cuttings and layers. The pistacio-nut or "green almond" (*P. vera*) is usually grown from seeds, which are planted where the trees are to stand. It is sometimes grafted upon *P. Terebinthus*, to give it greater vigor.

Pitcairnia. See Billbergia.

Pitcher-plant. See Nepenthes and Sarracenia.

Pittosporum. *Pittosporaceæ.*

Seeds, and by cuttings of the growing or ripe wood, under glass.

Planera (Planer-tree). *Urticaceæ.*

Propagated by seeds, which should be handled like elm seeds.

Plane-tree. See Platanus.

Plantago (Plantain). *Plantaginaceæ.*

Seeds. The perennial species also by division.

Plantain (fruit). See Banana; also Musa.

Platanus (Plane-tree, Buttonwood; Sycamore, improperly). *Platanaceæ.*

Usually propagated by seeds, but layers and ripe-wood cuttings may be employed.

Platycerium (Stag's-Horn Fern). *Filices.*

Chiefly by division. See Ferns.

Platycodon, Wahlenbergia. *Campanulaceæ.*

Propagated by seeds, and, when old plants are obtainable, by division.

Plectocomia. *Palmaceæ.*

Seeds. May be increased by suckers.

Plum (*Prunus*, many species). *Rosaceæ.*

There are so many species of plums in cultivation, and the varieties of the same species are often so different in constitution and habit, that it is difficult to give advice concerning their propagation. All the species grow readily from fresh, well-ripened seeds. The pits should be removed from the pulp and then stratified until spring. If they are allowed to freeze, the germination will be more uniform, as the pits will be more easily opened by the swelling embryo. Plum pits are rarely cracked by

Plum, continued.

hand. The strong-growing species and varieties, especially southwards, will give stocks strong enough to bud the first season; but the weaker ones must stand until the next season after the seeds are planted. In all the northern states, however, plum pits are usually sown in seed-beds, in the same manner as apple and pear seeds. The seedlings are taken up in the fall, and the following spring set out in nursery rows, where they are budded in August.

Plums are extensively grown from suckers, which spring in great numbers from the roots of many species. In France this method of propagation is largely used. So long as graftage does not intervene, the sprouts will reproduce the variety; and even in grafted or budded trees this sometimes occurs, but it is probably because the tree has become own-rooted from the rooting of the cion. It is a common notion that trees grown from suckers sprout or sucker worse than those grown from seeds. Layers are also sometimes employed for the propagation of the plum. Strong stools (page 39) are grown, and the long and strong shoots are covered in spring throughout their length—the tips only being exposed—and every bud will produce a plant. Strong shoots of vigorous sorts will give plants strong enough the first fall to be removed into nursery rows. Mound-layering is also employed with good results. Root cuttings, handled like those of blackberry, grow readily, but some growers suppose that they produce trees which sucker badly. Many plums grow readily from cuttings of the mature recent wood, treated the same as long grape cuttings. This is especially true of the Marianna (which is a form of Myrobalan, or a hybrid of it and some native plum of the Wild Goose type), which is grown almost exclusively from cuttings. Some sorts of the common garden plum (*P. domestica*) also grow from cuttings.

Plums are worked in various ways, but ordinary shield-budding is usually employed in late summer or early fall, as for peaches and cherries. Root-grafting by the common whip method is sometimes employed, especially when own-rooted trees are desired (pages 109, 110). In the north and east, the common plum (*P. domestica*) is habitually worked upon stocks of the same species, and these are always to be preferred. These stocks, if seedlings, are apt to be very variable in size and habit, and sometimes half or more of any batch, even from selected seeds, are practically worthless. Stocks from inferior or constant

Plum, continued.

varieties are, therefore, essential. Such stocks are largely imported; but there are some varieties which can be relied upon in this country. One of the best of these domestic stocks is the Horse plum, a small and purple-fruited variety of *Prunus domestica*, which gives very uniform seedlings. This is sometimes used in New York. It is simply a spontaneous or wilding plum, in thickets and along roadsides. The French stocks which are in most common use are St. Julien and Black Damas. The Myrobalan (*P. cerasifera*) is chiefly used for plums, however, because of its cheapness and the readiness with which all varieties take on it. The peach is often used as a plum stock, and it is valuable in the south, especially for light soils. In the north plum stocks are better. Marianna is used southwards, very likely too freely. Almond stocks, especially for the French prune and for light soils, are considerably used in California. The apricot is sometimes employed, but results appear to be poor or indifferent, on the whole. Prunes thrive upon the above stocks also.

Various stocks dwarf the plum. The chief dwarf stock at present is the Myrobalan. This is imported. It is easily grown from seeds, or sometimes from cuttings. Although the Myrobalan, like the Mahaleb cherry, is a slow grower, the dwarfing of the top depends more upon subsequent pruning than upon the root. The Mirabelle (*P. cerasifera*), a foreign stock, is sometimes used. The many species of native plums, of the *Prunus Americana* and *P. angustifolia* (Chickasaw) types, are good stocks for dwarf or intermediate trees. In most cases, the bud or graft grows luxuriantly for two or three years, and thereafter grows rather slowly. It is best to bud or graft low upon these stocks. Unless the tops are freely and persistently headed in, however, dwarf plum trees are not secured. The only exception to this statement seems to be in the use of the native dwarf cherry stocks (*Prunus pumila* and *P. Besseyi*), which have been used in an experimental way with much promise.

The native or American plums are budded upon native seedlings, or rarely upon *Prunus domestica* seedlings; or they are grown from cuttings, as in the case of Marianna.

The Japanese plums are worked upon peach, common plum, natives, or Marianna. Peach and Marianna are mostly used, but as the Japanese plums begin to bear freely their own seedlings will no doubt be used for stocks,

Plum, concluded.

and this may be expected to be an advantage. Peach is probably preferable to Marianna.

Prunus Simonii works upon peach, common plum, Myrobalan and Marianna, chiefly upon the first.

The ornamental plums are worked upon the same stocks as the fruit-bearing sorts. See Prunus.

Plums (like cherries) can be top-grafted the same as apples, but the cions must be kept completely dormant. It is preferable to graft very early in the spring.

Plum, Coco. See Chrysobalanus.

Plumbago (Leadwort). *Plumbaginaceæ.*

Seeds, division and cuttings. Cuttings are made from firm, nearly mature wood, and should be given mild bottom heat.

Podocarpus. *Coniferæ.*

Usually grown from cuttings of firm wood under cover.

Podophyllum (May Apple, Mandrake; erroneously Duck's Foot). *Berberidaceæ.*

Seeds (stratified or sown as soon as ripe) and division.

Poinciana. *Leguminosæ.*

Propagation by seeds.

Poinsettia. *Euphorbiaceæ.*

Cuttings of growing shoots, of two or three buds each, handled upon a cutting-bench or in a frame. Many propagators prefer to let the cuttings lie exposed two or three days before setting them. Cuttings of ripened wood can be used to good advantage where the heat is rather low. See Euphorbia.

Polemonium. *Polemoniaceæ.*

Propagated by seeds and by division.

Polianthes (Tuberose). *Amaryllidaceæ.*

Increased by bulbels. Remove these from the parent bulb in the fall, and keep in a warm, dry place until the following spring. The soil should be light, rich and moist throughout the summer. Before frost comes in the fall, take the bulbs up, and when dry, cut off the leaves. The bulbs should be kept as during the preceding winter, and the culture during the following year is the same as during the first. The bulbs usually flower the second or third summer.

Polyanthus. See Primula.

Polygala (Milkwort). *Polygalaceæ.*

Seeds; sometimes by division, and by cuttings of young shoots under cover, particularly for tropical species.

Polygonatum (Solomon's Seal). *Liliaceæ.*

Propagated by seeds and by division.

Polygonum (Knot-Grass or Knot-Weed). *Polygonaceæ.*

Seeds. The perennials are also easily increased by division of the rootstocks, and by cuttings. See Sacaline.

Polypodium (Polypody). *Filices.*

Division usually. See Ferns.

Pomegranate (*Punica Granatum*). *Lythraceæ.*

Largely by seeds, and all varieties are increased by cuttings, suckers, layers, and scarce sorts by grafting on a common sort.

Pomelo, Shaddock (*Citrus Decumana*). *Rutaceæ.*

Usually grown from seeds, but it may be budded upon pomelo or orange stocks, as in the Orange, which see.

Pontederia (Pickerel Weed). *Pontederiaceæ.*

Seeds rarely. Mostly by division. See, also, Eichhornia.

Poppy. See Papaver.

Populus (Poplar, Aspen, Cottonwood). *Salicaceæ.*

Seeds, sown as soon as ripe and raked in, in light soil. Suckers are also used. Most often increased by cuttings of ripe wood, taken in fall and spring. The weeping forms are stock-grafted upon upright sorts, chiefly upon *P. grandidentata.*

Portugal Laurel. See Prunus.

Portulaca (Purslane, Rose Moss). *Portulacaceæ.*

The annuals are raised from seed. Varieties are sometimes propagated by cuttings.

Potato (*Solanum tuberosum*). *Solanaceæ.*

Tubers, either whole or variously divided. Also rarely by stem cuttings. See page 60.

Potentilla, including Horkelia (Cinquefoil, **Five-Finger**), *Rosaceæ.*

Seeds, layers, division, green cuttings.

Poterium, including Sanguisorba (Burnet). *Rosaceæ.*

The herbaceous kinds are increased by seeds and division. The shrubs are raised from soft cuttings, under glass. See Burnet.

Prickly Ash. See Zanthoxylum.

Prickly Pear. See Opuntia.

Pride of India. See Melia.

Prim. See Ligustrum.

Primula, Polyanthus (Primrose, Cowslip). *Primulaceæ.*

Seeds, sown carefully in very fine soil, under glass. The seeds should be fresh; old ones often lie dormant a year. Many sorts are increased by division. See Auricula.

Prinos. See Ilex.

Pritchardia. *Palmaceæ.*

Increased by seeds.

Privet. See Ligustrum.

Prune. See Plum.

Prunus, Amygdalus. *Rosaceæ.*

The dwarf almonds (*Amygdalus*) are increased by seeds, division, cuttings, and by budding upon seedling plum or peach stocks; also by root cuttings. Peach stocks give larger trees at first than plum stocks, but the trees are not so long-lived. Perhaps ten years may be considered the average life of most ornamental almonds upon the peach, while upon the plum they may persist twenty-five years or more. (See Almond.) The ornamental cherries, peaches, etc., are propagated in essentially the same manner as the fruit-bearing varieties. *P. Lauro-Cerasus* and *P. Lusitanica*, the cherry laurel and Portugal laurel, may be propagated by short cuttings of ripened wood, in a cool greenhouse in autumn. *P. Pissardii* propagates by cuttings of the soft wood and, with more difficulty, from cuttings of dormant wood. Soft cuttings succeed well with many of the double-flowering plums and cherries, if the wood is grown under glass. See Apricot, Cherry, Peach, Plum.

Pseudotsuga. *Coniferæ.*

Propagated the same as Abies, which see.

Psidium. See Guava.

Psoralea *Leguminosæ.*

Seeds, divisions and cuttings of growing shoots, placed under glass. The tubiferous species, as the "pomme blanche" or Indian potato (*P. esculenta*) are increased by tubers or divisions of them.

Ptelea (Hop-tree). *Rutaceæ.*

Increased by seeds, sown in autumn or stratified, or by layers. The varieties may be grafted on the common forms.

Pteris (Brake, Bracken). *Filices.*

Easily grown from spores. See Ferns.

Pterocarya. *Juglandaceæ.*

Increased by seeds, suckers and layers.

Ptychosperma, Seaforthia (Australia Feather-palm). *Palmaceæ.*

Seeds in heat.

Pulmonaria See Mertensia.

Pumpkin (*Cucurbita*, three species). *Cucurbitaceæ.*

Seeds, when the weather is settled.

Punica. See Pomegranate.

Puschkinia, Adamsia. *Liliaceæ.*

Increased by dividing the bulbs, which should be done every two or three years.

Pyrethrum. See Chrysanthemum.

Pyrola (Shin-leaf, Wintergreen). *Ericaceæ.*

Propagated by division ; very rarely from seeds.

Pyrus. *Rosaceæ.*

The ornamental species and varieties of apples and crabs are budded or grafted upon common apple stocks. The mountain ashes are grown from stratified seeds, which usually lie dormant until the second year, or the varieties are budded or grafted upon stocks of the common species (*P. Aucuparia*). Layers and green cuttings are occasionally employed for various species and varieties of pyrus. See also Apple, Pear, Quince. It is a good plan to obtain stocks as nearly related to the plant which is to be propagated as possible; *e. g.*, Parkman's pyrus does better on *P. floribunda* than on the common apple stock. The wild crabs can be worked upon the apple when stocks of their own species cannot be had.

Quamoclit. See Ipomœa.

Quassia. *Simarubaceæ.*

Cuttings of ripe shoots under glass.

Quercus (Oak). *Cupuliferæ.*

Stocks are grown readily from seeds, which may be sown in the fall without stratification. Take care that vermin do not dig up the acorns. The evergreen species are sometimes grown from cuttings. Varieties are grafted on stocks grown from wild acorns. The stocks are potted in the fall, and the grafting (generally the veneer-graft) is performed in January and February, or sometimes in August.

Quince (*Pyrus Cydonia*, *P. Cathayensis*, *P. Japonica*, etc.). *Rosaceæ.*

All quinces can be grown from seeds, the same as apples and pears; but seeds are not common in the market, and are, therefore, little used. The fruit-bearing quinces are propagated most cheaply by means of cuttings of mature wood or by mound-layering. Cuttings are taken in the fall, and are stored in sand, moss or sawdust until spring, when they are planted outdoors. Long cuttings—10 to 12 inches—are usually most successful, as they reach into uniformly moist earth. Cuttings are usually made of the recent wood, and preferably with a heel, but wood two or three years old will usually grow. With some varieties and upon some soils, there is considerable uncertainty, and layerage is therefore often employed. Mound-layering (see page 39) is practiced where extra strong plants are required. Long root-cuttings, treated like those of the blackberry and raspberry, will also grow. Many nurserymen bud- or root-graft the better varieties upon stocks of Angers or other strong sorts. These stocks are imported from Europe (and are the same as those used for dwarf pears). These imported plants are grown both from cuttings and mound-layers, the greater part of them from the latter, but seeds are occasionally employed. These stocks are two years old when imported, having been transplanted the first year from the cutting-bed or the stool-yard. In order to secure extra strong plants and a uniform stand, some growers graft quince cuttings upon pieces of apple or pear roots. In such cases the plants should be taken up in the fall, when the quince will be found to have sent out roots of its own; the apple sprouts (or even the entire root) should be removed, and the quince replanted the following spring in the nursery row, otherwise suckers frequently spring from the stock and interfere with the growth of the quince. The union is sufficient to nurse the cion for two or three years.

The flowering or Japanese quince is best propagated by short root-cuttings, which are usually made in the fall, and scattered in drills in frames or in a well-prepared border in spring. Cuttings of firm, nearly mature wood, handled in frames, will grow, but they are not often used. The double varieties are root-grafted upon common stocks of *P. Japonica* in winter. The plants are then grown on in pots. Common quince (*P. Cydonia*) stocks are occasionally used, but they are not in favor. The Chinese quince (*Pyrus Cathayensis*) is worked upon the common quince.

Radish (*Raphanus sativus*). *Cruciferæ.*

Seeds, usually sown where the plants are to grow

Ragged Robin. See Lychnis.

Ramondia, Myconia. *Gesneraceæ.*

Propagated by seeds or division.

Rampion (*Campanula Rapunculus*). *Campanulaceæ.*

Seeds, where the plants are to stand.

Ranunculus (Buttercup, Crowfoot). *Ranunculaceæ.*

Propagated by seeds and by division.

Raphia. *Palmaceæ.*

Seeds.

Raspberry (*Rubus strigosus, R. occidentalis,* etc.). *Rosaceæ.*

New varieties are obtained from seeds, which are washed from the pulp and sown immediately, or stratified. The black-cap varieties are grown mostly from root-tips, as described on page 36. If the ground is loose and mellow, the tips will commonly take root themselves, but upon hard ground the tip may have to be held in place by a stone or clod. Some strong-growing varieties, like the Gregg, especially in windy localities, have to be held down. The red varieties increase rapidly by means of suckers which spring up from the roots. Better plants are obtained by means of root cuttings, however, as described under Blackberry (see also Fig. 62). Black-caps may be increased by root cuttings. These cuttings are best handled in warm coldframes or mild hotbeds, being planted very early in spring. By the time the weather is settled, they will be large enough to plant in nursery rows.

Red-bud. See Cercis.

Red Cedar. See Juniperus.

Reinwardtia, Linum in part, of gardeners. *Linaceæ.*

Seeds. Cuttings of strong shoots in heat.

Renanthera. See Ærides.

Reseda (Mignonette). *Resedaceæ.*

Seeds. For winter flowering, seeds are sown in July. Also grown from cuttings.

Resurrection Plant. See Anastatica.

Retinospora, species of *Chamæcyparis* (Japanese Arbor-Vitæ). *Coniferæ.*

Grown sometimes from seeds, which should be denuded of pulp. Layers of tender branches are sometimes employed. Most commonly grown from cuttings. These are made from tips of growing or ripened shoots, and are 2 or 3 inches long, with all the leaves left on. They are usually, from necessity, variously branched. The soft cuttings are usually taken from forced plants, and are handled in a close frame or under a bell-glass, with bottom heat. In commercial establishments the cuttings of ripe wood are preferred. The following is the practice of one of the largest nurseries in the country: Cuttings of the entire season's growth, cut to a heel, are taken in October and November, and are placed in sand in boxes in gentle heat, as in a propagating-house. By February the roots will be formed, and the boxes are then placed in a cool house where the temperature is about 50°. Early in spring (about April 1st) the boxes are placed outdoors in coldframes, where they remain until May, until frost is over. The boxes are then removed from the frames and are set on boards in a shady place, where they are left until fall. In the fall—having been nearly a year in the boxes—the plants are shaken out and are heeled-in in a cellar. The next spring they are planted out in beds, and during the following summer and winter they are given some protection from sun and cold. Yews and arbor-vitæs are handled in the same way.

Retinosporas are often grafted upon retinospora or common arbor-vitæ stocks. This operation is usually performed upon potted plants in winter by the veneer method.

Rhamnus, including Frangula (Buckthorn). *Rhamnaceæ.*

The hardy kinds may be increased by means of seeds or by layers. The stove and greenhouse species may be multiplied by cuttings of growing parts. Seeds should be stratified.

Rheum (Rhubarb, Pie-plant, Wine-plant). *Polygonaceæ.*

Increased by seeds and by division. Each division should contain at least one bud on the crown. Seeds may be sown where the plants are to stand, but will not reproduce the varieties, and three years are required for the plants to mature.

Rhipsalis, including Lepismium, Pfeiffera. *Cactaceæ.*

Cuttings, after having been dried for a few days, should be inserted in coarse gravel or sand. See Cactus.

Rhodanthe. See Helipterum.

Rhododendron, Azalea (Rose-Bay). *Ericaceæ.*

Seeds are largely employed, but they are small and light, and must be carefully handled. They are sown in spring in pans or boxes in a soil or sandy peat, care being taken to cover them very lightly and not to dislodge them when applying water. They are handled in coldframes or in a cool house, and the young plants must be shaded. The plants are commonly allowed to remain a year in the boxes. Low-growing plants are often layered. Cuttings of growing wood, cut to a heel, are sometimes employed, being made in summer and handled in a frame, but the percentage of rooted plants will often be small. Rhododendrons are extensively grafted, the veneer method being most used. The operation is performed upon potted plants in late summer or early fall, or sometimes in a cool house in early spring. Most of the leaves are allowed to remain upon the cion. The plants are then placed in densely shaded cool frames (Fig. 47), and are nearly covered with sphagnum. Various stocks are employed, but for severe climates the hardy species, like *R. Catawbiense* and *R. maximum*, are probably best. *R. Ponticum* is extensively used in Europe, but it is not hardy enough for the north, unless worked low and planted deep. See Azalea.

Rhodotypos. Seeds; or like Kerria.

Rhubarb. See Rheum.

Rhus (Sumach). *Anacardiaceæ.*

Seeds, layers, suckers, root cuttings, and cuttings of green or ripe wood. Suckers are oftenest used.

Rhynchospermum. See Trachelospermum.

Ribes (Currant, Gooseberry). *Saxifragaceæ.*

Seeds, which should be sown as soon as ripe, or else stratified for new varieties. Commonly from ripe cuttings. See Currant and Gooseberry.

Richardia (Calla). *Aroideæ.*

Offsets, which should be removed and potted off when the plants are at rest. Old crowns may be divided.

Ricinus (Castor Bean). *Euphorbiaceæ.*

Seeds, which in the north are started indoors.

Rivina (Hoop Withy). *Phytolaccaceæ.*

Readily propagated by seeds; also by cuttings, inserted during spring in heat.

Robinia (Locust, Rose Acacia). *Leguminosæ.*

Seeds, sown in fall or spring, and which usually germinate better if soaked in hot water previous to sowing. Also grown from layers and root cuttings. Named varieties are grafted or budded, the common locust stock (*R. Pseudacacia*) being preferred, even for the rose acacia (*R. hispida*).

Rocambole (*Allium Scorodoprasum*). *Liliaceæ.*

"Cloves," or division of the bulb.

Rocket, ornamental sorts. See Hesperis.

Rocket Salad (*Eruca sativa*). *Cruciferæ.*

Seeds, sown where the plants are to grow.

Rock-Rose. See Cistus.

Romneya. *Papaveraceæ.*

Propagated by seeds in spring.

Rosa (Rose). *Rosaceæ.*

New varieties, and sometimes stocks, are grown from seeds, which are sown as soon as ripe, or kept in the hips until spring. The hardy kinds are usually sown in well prepared beds outdoors. Roses are sometimes grown from layers, and often from root cuttings, after the manner of blackberries. The common way of propagating roses, however, is by means of short cuttings of firm or nearly mature wood, handled under glass, with a mild bottom heat (65° or 70°). They are commonly made in February or March from forced plants. The cuttings are made in various fashions, some persons allowing most of the leaves to remain, and some preferring to cut most of them off, as in Fig. 74. They are commonly cut to one-bud lengths, like Fig. 76. Long cuttings of ripened wood, handled in a cool greenhouse or in frames, may also be employed for the various perpetual and climbing roses. Most growers feel that the best plants are obtained from cuttings, but most varieties do well when budded upon congenial and strong stocks. Budding by the common shield method is considerably employed, and veneer-grafting is sometimes used. The stocks are grown either from seeds or cuttings. A common stock is the manetti, which is a strong and hardy type. The eyes should be cut out of the manetti stock below the bud, to avoid sprouting. Because the manetti suckers badly, various wild briars are much used in Europe. The bud is often inserted 2 to 4 feet high, making "standard" roses. These are practically unknown in this country, except as sparingly imported. The multiflora rose is also a good stock, especially for early results. These manetti and multiflora stocks (and some others) are imported from Europe as yearling cuttings. For outdoor propagating, they are "dressed" much like apple stocks (Fig. 86), and are budded the year in which they are planted in the nursery row. The gardener may grow his own stocks of these (particularly of multiflora) from hard-wood cuttings made in spring, and these cuttings should be fit for working in the following fall and winter. Home-grown seedlings should be two years old (unless very strong) before they are budded. Hybrid perpetual roses make excellent pot plants in a short time when winter grafted, with dormant wood, upon multiflora stocks. A stock somewhat used about Boston for some of the hybrid perpetuals, with excellent results, is *Rosa Watsoniana*, a Japanese species. This is a slender stock, and is grafted, not budded. "Worked" roses

Rosa, concluded.

are in greater favor in Europe than in this country, and our various native roses have, therefore, received little attention as stocks. The common sweet briar of the roadsides (which is an introduced species) is sometimes used for stocks. *R. Wichuriana* is easily propagated by long cuttings of year-old wood in the open air.

Rosemary (*Rosmarinus officinalis*). *Labiatæ*.

Seeds and division.

Rubber-plant. See Ficus (*F. elastica*).

Rubus (Bramble). *Rosaceæ*.

Seeds, which should be stratified or sown as soon as ripe. Root cuttings and suckers are mostly employed. The seeds of *R. deliciosus* require two years for germination. See Blackberry, Dewberry, Raspberry, Wineberry.

Rudbeckia, including Lepachys, Obeliscaria (Cone Flower). *Compositæ*.

Propagated by seeds or division.

Rue. See Ruta.

Ruscus (Butcher's Broom, Alexandrian Laurel). *Liliaceæ*.

Root suckers. Also seeds, when obtainable.

Rush. See Juncus.

Russelia. *Scrophulariaceæ*.

Seeds. Green cuttings under glass is the common method.

Ruta (Herb of Grace, Rue). *Rutaceæ*.

Propagated by seeds, division and cuttings. Meadow Rue, see Thalictrum.

Sabal (Palmetto). *Palmaceæ*.

Propagated by seeds, and by suckers, which should be taken when about one foot long. If they have no roots they must be carefully handled.

Sabbatia (American Centaury). *Gentianaceæ*.

May be raised from seeds, which should be sown thinly in pans, or in a shady border. Division of old plants.

Sacaline, or **Saghalin** (*Polygonum Sachalinense*). *Polygonaceæ*.

Division of the roots (*i. e.*, root cuttings) into small pieces.

Saffron (*Carthamus tinctorius*). *Compositæ*.

Propagated by seeds, in open air in spring. Saffron is also *Crocus sativus*. See Crocus.

Sage (*Salvia officinalis*). *Labiatæ*.

Seeds, sown in spring where the plants are to stand. Also by division, but seeds give better plants. Sage plantations should be renewed every two or three years. Good plants may be grown from cuttings. See Salvia.

Sage Palm. See Cycas.

Saintfoin. See Onobrychis.

Saint John's Bread. See Carob.

Saintpaulia. *Gesneraceæ*.

Grown easily from seeds, sown on the surface. Also from leaf cuttings. Handled like Sinningia, except that it is not tuber-bearing.

Salisburia. See Ginkgo.

Salix (Willow, Osier, Sallow). *Salicaceæ*.

All the willows grow readily from cuttings of ripe wood of almost any age. The low and weeping varieties are top-worked upon any common upright stocks. Kilmarnock (weeping form of *Salix Caprea*), Rosmarinifolia (*S. incana*), and other named varieties are worked upon cutting-grown stocks of *S. Caprea*.

Salpiglossis. *Scrophulariaceæ*.

Propagated by seeds in open air, or they may be started under glass.

Salsify (*Tragopogon porrifolius*). *Compositæ*.

Seeds, sown in spring where the plants are to remain.

Salvia, including Sclarea (Sage). *Labiatæ*.

May be increased by seeds, sown thinly and placed in a little warmth. Also by cuttings; these will root readily in heat, if they are rather soft and in a growing state. See Sage.

Sambucus (Elder). *Caprifoliaceæ*.

Seeds, handled like those of raspberries and blackberries. Named kinds are grown from cuttings of mature wood, and by layers.

Sandoricum (Sandal-tree). *Meliaceæ.*

Seeds. Cuttings, in sand under glass, in heat.

Sanguinaria (Blood-root, Red Puccoon). *Papaveraceæ.*

Propagated by means of seeds, or (more commonly) by division of the rootstocks.

Sanguisorba. See Poterium.

Sansevieria, Salmia (Bowstring Hemp). *Hæmodoraceæ.*

Young plants are obtained from suckers.

Sapodilla, or **Sapodilla Plum.** See Sapota.

Saponaria, including Vaccaria (Bouncing Bet, Fullers' Herb, Soapwort). *Caryophyllaceæ.*

Increased by seeds and by division. The hardy annual and biennial kinds may be simply sown in the open border.

Sapota, Achras. *Sapotaceæ.*

Seeds and cuttings. In tropical countries the sapodilla (*S. Achras*) is raised entirely from seeds.

Sarracenia (Indian Cup, Pitcher Plant, Side-saddle Flower, Trumpet Leaf). *Sarraceniaceæ.*

Increased by dividing the crowns. Sometimes by seeds, sown in moss in a cool frame.

Sassafras. *Lauraceæ.*

Increased by seeds, suckers and root cuttings.

Satyrium. *Orchidaceæ.*

Division of the plants, as new growth is commencing. See also under Orchids.

Sauromatum. *Aroideæ.*

Increased by offsets.

Savin. See Juniperus.

Savory (*Satureia hortensis, S. montana*). *Labiatæ.*

Seeds, sown where the plants are to remain; or the winter savory (*S. montana*), which is a perennial; also by division.

Savoy. See Cabbage.

Saxifraga (Saxifrage, Rockfoil). *Saxifragaceæ.*

Seeds, divisions, and in some species (as *S. sarmentosa*, the "strawberry geranium") by runners.

Scabiosa (Mourning Bride, Pin-cushion Flower). *Dipsaceæ.*

Seeds, usually sown in the open, and sometimes by division.

Scævola. *Goodenovieæ.*

Seeds. Cuttings should be inserted in a compost of peat and sand, under glass.

Schinus (Pepper-tree, of California). *Anacardiaceæ.*

Propagated by seeds. Cuttings, in greenhouses.

Schismatoglottis. *Aroideæ.*

Increased by division.

Schizandra. *Magnoliaceæ.*

Seeds, when procurable. Propagation is effected by layers; by ripened cuttings, which should be inserted in sand under glass.

Schizanthus (Butterfly, or Fringe Flower). *Solanaceæ.*

The half-hardy kinds are increased by seeds sown in a little heat in spring. The seed of the hardy sorts may be sown in the open ground in early spring.

Schizostylis. *Iridaceæ.*

Propagated by seeds and by division.

Sciadophyllum. *Araliaceæ.*

Seeds. Cuttings, in sand under glass, in moderate heat.

Sciadopitys (Umbrella Pine). *Coniferæ.*

Slowly propagated by imported seeds. But cuttings of the half-ripened shoots, taken off in summer and inserted in sand, in heat, root readily.

Scilla (Squill, Wild Hyacinth). *Liliaceæ.*

Slowly increased by seeds, but usually by bulbels.

Scirpus Tubernæmontanus of florists, is Juncus, which see.

Scolopendrium. See Ferns.

Scorzonera (Black Salsify). *Compositæ.*

Seeds, sown where the plants are to stand.

Scotch Broom. See Cytisus.

Screw Pine. See Pandanus.

Seaforthia. See Ptychosperma.

Sea-kale (*Crămbe maritima*). *Cruciferæ.*

Seeds, sown without being shelled, usually in a seed-bed. When the young plants have made three or four leaves, they should be removed to permanent quarters. Seedlings should furnish crops in three years. By root-cuttings, four or five inches long, taken from well established plants. These should give plants strong enough for cutting in two years.

Seaside Grape. See Coccoloba.

Sechium (Choko). *Cucurbitaceæ.*

Seeds. Root-tubers.

Sedge. See Carex.

Sedum (Orpine, Stonecrop). *Crassulaceæ.*

Propagation may be effected by seeds, by division of the tufts, by cuttings of stems or leaves in spring.

Selaginella. *Lycopodiaceæ.*

Spores, as for Ferns (which see). Short cuttings, inserted in early spring, in pots or pans.

Sempervivum (House Leek). *Crassulaceæ.*

Readily increased by seeds, or by the young plants which appear around the old one at the base.

Senecio, including Cacalia, Farfugium, Jacobæa, Ligularia (Grounsel, Ragweed). *Compositæ.*

The annuals are propagated by seeds. Others may be increased by seeds, by division, or by cuttings of both the roots and shoots. German Ivy (*Senecio scandens*) is easily multiplied by cuttings of the running shoots.

Sensitive Plant. See Mimosa.

Sequoia, Wellingtonia (Redwood). *Coniferæ.*

Seeds, which must be handled in a frame or half-shady place. Layers, and cuttings handled like those of retinospora and yew.

Sericographis. See Justicia.

Service-berry. See Amelanchier.

Sesamum (Bene). *Pedalineæ.*

Seeds, sown under glass, or in the south in the open border.

Sesbania (Pea-tree). *Leguminosæ.*

Seeds for annual species; the shrubby kinds by cuttings of the half-ripened shoots under glass, in heat.

Shad-bush. See Amelanchier.

Shaddock. See Pomelo.

Shallot (*Allium Ascalonicum*). *Liliaceæ.*

Grown from "cloves," which are formed by the breaking up of the main bulb.

Shell-bark Hickory (Shag-bark). See Hicoria.

Shepherdia (Buffalo Berry). *Elæagnaceæ.*

Increased by seeds sown in the fall or stratified until spring.

Sibbaldia. See Potentilla.

Siberian Pea-tree. See Caragana.

Side-saddle Flower. See Sarracenia.

Silene (Campion, Catchfly). *Caryophyllaceæ.*

By seeds, division, and cuttings.

Silk-cotton Tree. See Bombax.

Silphium (Rosin-plant, Compass-plant). *Compositæ.*

Propagated by seeds and by division.

Silver Bell. See Halesia.

Sinningia (Gloxinia.) *Gesneraceæ.*

Seeds should be sown the latter part of winter, in well-drained pots or small pans of finely sifted soil, of peat, leaf-mold and sand in about equal proportions. The seeds should be sown thinly and covered slightly, then carefully watered, and placed in a temperature of about 70° and kept shaded. Cuttings of the shoots may be taken when the old tubers are starting in spring, and placed in a close propagating frame. Leaf cuttings, with a small portion of the petiole attached, give excellent results, especially when the leaves are firm and nearly matured. Leaf cuttings are made after the fashion of Fig 81. A little tuber forms on the end of the leaf-stalk, and this is removed and handled like any small tuber. Also grafted on tubers (see page 129).

Sisyrinchium (Blue-eyed Grass, Satin Flower). *Iridaceæ.*

It may be increased in spring by seeds or by division.

Skimmia. *Rutaceæ.*

Seeds, in a frame. Also by layers, and by firm cuttings in gentle heat.

Skirret (*Sium Sisarum*). *Umbelliferæ.*

Seeds, offsets, or division.

Slipperwort. See Campanula and Calceolaria.

Smilacina (False Solomon's Seal). *Liliaceæ.*

Seeds. Division of roots.

Smilax (Green-Briar, American China Root). *Liliaceæ.*

Young plants are obtained by seeds, by layers, and by division of the root.

For the "Smilax" or Boston-vine of conservatories, see Myrsiphyllum.

Snapdragon. See Antirrhinum.

Snowball. See Viburnum.

Snowberry. See Symphoricarpus and Chiococca.

Snowdrop. See Galanthus.

Snowflake. See Leucoium.

Soapwort. See Saponaria.

Solandra. *Solanaceæ.*

Increased by seeds sown in spring; by cuttings, inserted in mold or tan. If small flowering plants are desired, the cuttings should be taken from flowering shoots.

Solanum (Nightshade). *Solanaceæ.*

The annuals, and most of the other species, are raised from seeds. The tuberous kinds may be increased by tubers, or division of them. The stove and greenhouse shrubby plants may be propagated by cuttings, inserted when young in a warm frame.

Soldanella. *Primulaceæ.*

Increased by seeds and by division.

Solea. See Ionidium.

Solidago (Golden Rod). *Compositæ.*

Seeds, sown in fall or spring, and by division.

Solomon's Seal. See Polygonatum.

Sonerila. *Melastomaceæ.*

Propagated by seeds; or by cuttings, which should be inserted singly in small pots during spring and placed in a frame in a propagating house.

Sophora. *Leguminosæ.*

Seeds, layers and cuttings of either ripened or growing wood. The named varieties are grafted upon common stocks.

Sorghum. *Gramineæ.*

Usually by seeds. Sometimes by cuttings, as in Sugar Cane, which see.

Sorrel (*Rumex*, several species). *Polygonaceæ.*

Seeds and division.

Sorrel-tree. See Oxydendrum.

Sour Gum. See Nyssa.

Spanish Bayonet. See Yucca.

Sparaxis. *Iridaceæ.*

Usually by offsets. Seeds.

Sparmannia. *Tiliaceæ.*

Propagated by cuttings of half-ripened wood in spring.

Spathiphyllum, including Amomophyllum. *Aroideæ.*

Propagated sometimes by seeds sown in heat, but mostly by division of the rootstocks.

Spearmint (*Mentha viridis*). *Labiatæ.*

Commonly grown from cuttings of the creeping rootstocks.

Speedwell. See Veronica.

Sphæralcea (Globe Mallow). *Malvaceæ.*

Seeds; by cuttings of the young growth under glass, and kept shaded until rooted.

Spice-bush. See Lindera.

Spiderwort. See Tradescantia.

Spinage (*Spinacia oleracea*). *Chenopodiaceæ.*

Seeds, sown usually where the crop is to stand, either in fall or spring.

Spiræa (Spirea, Meadow-Sweet). *Rosaceæ.*

Seeds, sown as soon as ripe or stratified until spring. Commonly increased by cuttings, either of mature or green wood. Green cuttings usually make the best plants. These are made in summer and handled in frames. Some sorts, as *S. ariæfolia*, *S. opulifolia* and varieties (Nine-bark, now known as *Neillia* or *Physocarpus opulifolia*) and *S. prunifolia*, are usually grown from layers put down in spring. The herbaceous kinds are often increased by division. Plants forced in winter give excellent cutting-wood, which should be taken when the growth is completed.

Spondias (Hog Plum, Otaheite Apple or Plum). *Anacardiaceæ.*

Seeds; by large cuttings of growing wood, which should be inserted in sand or mold, in heat.

Spruce. See Picea and Abies.

Squash (*Cucurbita*, three species). *Cucurbitaceæ.*

Seeds, when the weather beoomes warm.

Squill. See Scilla.

Stachys, Betonica, Galeopsis (Hedge Nettle, Woundwort). *Labiatæ.*

Seeds, divisions, or cuttings. Some species (as the Crosnes, Chorogi, or "*S. tuberifera*" of recent introduction) are increased by subterranean tubers.

Staff-tree. See Celastrus.

Stanhopea. *Orchidaceæ.*

Division of the old roots. See also under Orchids.

Stapelia (Carrion Flower). *Asclepiadaceæ.*

Seeds; commonly by cuttings in heat.

Staphylea (Bladder-nut). *Sapindaceæ.*

Seeds, sown as soon as ripe or stratified until spring. By suckers, layers, and cuttings of roots or of mature wood.

Star Apple. See Chrysophyllum.

Star of Bethlehem. See Ornithogalum.

Statice (Sea Lavender, See Pink). *Plumbaginaceæ.*

The annuals and biennials may be increased by seeds sown in early spring, in a frame. The perennials by seeds, or by carefully made divisions. Greenhouse species should be propagated by cuttings inserted in small single pots during early spring, and placed under glass.

Staurostigma. *Aroideæ.*

Seeds sown in bottom heat; or by division of the tubers.

Stephanotis, Jasminanthes. *Asclepiadaceæ.*

Propagated by seeds; also by cuttings of the previous year's growth inserted singly in pots, in spring, and placed in a close frame with a temperature of 60°.

Sterculia. *Sterculiaceæ.*

Seeds. Increased by ripened cuttings, which should be taken with the leaves on, and placed under glass. Those of the stove species should be placed in a moist heat.

Stevia. See Piqueria.

Stigmaphyllon. *Malpighiaceæ.*

Seeds. Cuttings of ripened wood, inserted in sandy soil under glass, in heat.

Stillingia. *Euphorbiaceæ.*

Easily propagated by imported seeds.

Stock. See Matthiola.

Stokesia. *Compositæ.*

Propagation by seeds and by division.

Stonecrop. See Sedum.

Strawberry (*Fragaria*). *Rosaceæ.*

New sorts are grown from seeds, which are usually sown as soon as ripe; or they may be kept until the following spring, either dry or in stratification. Varieties are commonly increased by offsets, or plants formed at the joints of runners. These runners appear after the fruit is off. If strong plants are desired, the runner should be headed-in, and only one plant allowed to form on each runner. The ground should be soft and somewhat moist, to enable the young plants to obtain a foothold. Plants strong enough for setting are obtained in August and September of the same year in which they start. Ordinarily, the runners will take root without artificial aid; but in hard soils, or with new or scarce varieties, the joints are sometimes held down with a pebble or bit of earth. New varieties are often propagated throughout the season from plants which are highly cultivated, and which are not allowed to fruit. Very strong plants are obtained by growing them in pots. A 3-inch pot is sunk below the runner, and the joint is held upon it by a stone or clod. The runner is then pinched off, to prevent further growth, and to throw all its energy into the one plant. The pot should be filled with soft, rich earth. Shouldered pots are best, because they can be raised more easily than others, by catching the spade or trowel under the shoulder. The plants will fill the pots in three or four weeks, if the weather is favorable. Old tin fruit cans, which have been heated to remove the bottoms, can also be used.

Cuttings of the tips of runners are sometimes made and handled in a frame, as an additional means of rapidly increasing new kinds. These cuttings are really the castaway tips left from the headings-in or checking of the runners.

Strawberry Geranium. See Saxifraga.

Strawberry Tree. See Arbutus.

Strelitzia (Bird of Paradise Flower, Bird's-tongue Flower). *Scitamineæ.*

Increased by seeds, which should be sown in light soil, and the pots plunged in moist bottom heat. Also increased by suckers and by division of the old plants.

Streptocarpus (Cape Primrose). *Gesneraceæ.*

Readily propagated by seeds or by division.

Strobilanthes, including Goldfussia (Cone Head). *Acanthaceæ.*

Seeds. Cuttings, in any light soil under glass, in heat.

Struthiola. *Thymelæaceæ.*

Seeds, when obtainable. Cuttings in sand under a frame.

Stuartia. *Ternstrœmiaceæ.*

May be increased by seeds and layers, or by means of ripened cuttings, inserted in sand under a hand-glass. Seeds are oftenest used, where obtainable.

Styrax (Storax). *Styracaceæ*.

Seeds, which must be stratified, or else sown as soon as ripe. They usually lie dormant the first year. Also by layers and cuttings of green wood. Can be grafted upon other storaxes, or upon *Halesia tetraptera*.

Sugar Cane (*Saccharum officinarum*). *Gramineæ*.

Cuttings of the stems. The cuttings should possess a node or joint which bears one or more good buds. These cuttings are planted directly in the field, and the plants will reach maturity in two or three months. Propagation by seeds has been supposed to be impossible, but recent experiments at Kew indicate that it can be done.

Sumach. See Rhus.

Sundew. See Drosera.

Sunflower. See Helianthus.

Sun Rose. See Helianthemum.

Swainsona. *Leguminosæ*.

Seeds. Green cuttings under cover.

Swan River Daisy. See Brachycome.

Sweet Brier. See Rosa.

Sweet Cicely. See Myrrhis.

Sweet Pea. See Lathyrus.

Sweet Potato (*Ipomœa Batatas*). *Convolvulaceæ*.

Sweet potato plants are grown in hotbeds, coldframes or forcing houses (depending upon the latitude) from sound tubers of medium size. The tuber is laid upon a sandy or other loose bed, and is then covered with sand or sandy loam to a depth of 1 or 2 inches. Sometimes, to guard against rot, the tubers are not covered until the sprouts begin to appear. The tubers may be laid thickly upon the bed, but they are less apt to rot if they do not touch each other. Sometimes the tubers are cut in two lengthwise, the cut surface being placed down, in order to place all the plant-giving surface uppermost. In four or five weeks the young plants—3 to 5 inches high—are pulled off and planted, and others soon arise to take their places. One hand should be held firmly upon the soil over the tuber, while the sprout is pulled off, to keep it in place. Three or four crops of sprouts may be obtained from each tuber.

Sweet William (*Dianthus barbatus*). *Caryophyllaceæ.*

Seeds, sown indoors or in the border. Division of the plants. Best results are obtained by starting new seedlings every other year. See Dianthus.

Sycamore. See Platanus.

Symphoricarpus (Waxberry, St. Peter's Wort, Snowberry-tree, Indian Currant). *Caprifoliaceæ.*

Seeds, handled like those of blackberries. Also by suckers and cuttings.

Symphytum (Comfrey). *Borraginaceæ.*

May be increased by seeds and by division. Also easily by root cuttings.

Symplocos, including Hopea. *Styracaceæ.*

Seeds. Cuttings, in sand under glass.

Syringa (Lilac). *Oleaceæ.*

New varieties and stocks are grown from seeds, which are usually stratified until spring. Green cuttings, handled in frames in summer, are largely used. Cuttings of mature wood will grow; also cuttings of the roots. Layers and suckers are often employed. Varieties are extensively grafted or budded upon privet (Ligustrum) and common lilacs. Flute-budding is occasionally employed. Lilacs will grow for a time when worked upon the ash. Grafting succeeds well when performed in the open air.

Tabernæmontana. *Apocynaceæ.*

Increased by green cuttings, under glass, in moist heat.

Tacca, Ataccia. *Taccaceæ.*

Seeds, and division of the roots.

Tacsonia. See Passiflora.

Tagetes (Marigold). *Compositæ.*

Seeds, sown either indoors or out.

Tamarack. See Larix.

Tamarindus (Tamarind). *Leguminosæ.*

Young plants may be obtained from seeds sown in a hotbed, or outdoors in tropical countries. Cuttings, in sand under glass, in heat.

Tamarix, Tamarisk. *Tamariscineæ.*

Increased by ripe cuttings under glass, the greenhouse kinds in heat.

Tansy (*Tanacetum vulgare*). *Compositæ.*

Seeds and division.

Taro. Root tubers. See Caladium.

Tarragon (*Artemisia Dracunculus*). *Compositæ.*

A perennial herb, multiplied chiefly by division. Seeds may be used, if fresh.

Taxodium, Glyptostrobus (Bald Cypress). *Coniferæ.*

Seeds are usually employed. Layers. Cuttings of young wood in wet sand, or even water, under cover. The varieties of glyptostrobus may be veneer-grafted in August or September on *T. distichum.*

Taxus (Yew). *Coniferæ.*

Seeds, sown when gathered or else stratified. Layers. Cuttings of green wood under glass in summer, or of mature wood, as recommended for retinospora. The named varieties are veneer-grafted in August or early fall upon the upright kinds.

Tecoma (Trumpet-Creeper). *Bignoniaceæ.*

Seeds, layers, cuttings of firm shoots, but most commonly by root cuttings.

Terminalia (Tropical Almond). *Combretaceæ.*

Seeds; also by cuttings of green wood under glass.

Ternstrœmia. *Ternstrœmiaceæ.*

Seeds. Cuttings of the half-ripened shoots under glass, in bottom heat.

Testudinaria (Elephant's Foot). *Dioscoreaceæ.*

Grown from imported roots or seeds.

Teucrium (Germander). *Labiatæ.*

Seeds, division, and the shrubby kinds by cuttings under cover.

Thalictrum (Meadow Rue). *Ranunculaceæ.*

Propagated by seeds, and division. The varieties by cuttings.

Thea. See Camellia.

Theobroma (Cacao, Chocolate-tree). *Sterculiaceæ.*

Propagated by ripened cuttings, which should be placed in sand under glass, in heat.

Thrift. See Armeria.

Thrinax. *Palmaceæ.*

Seeds in heat.

Thunbergia. *Acanthaceæ.*

Seeds. Also cuttings of firm wood in a frame, for perennials.

Thunia. *Orchidaceæ.*

As the form of the pseudo-bulbs suggests, this genus is easily propagated by cuttings. These are made about 6 inches long and inserted in pots of sand. After standing in an ordinary propagating frame or moist stove for a short time, young growths will appear at the nodes. When large enough they are taken up and potted in ordinary compost. Two years, at least, are needed for them to attain to flowering size, but this is the best method where a large number of plants are wanted. See also under Orchids.

Thuya, including Biota (Arbor-Vitæ, White Cedar erroneously). *Coniferæ.*

Seeds, which should be gathered as soon as ripe (in the fall) and stratified or sown at once. Shade the seedlings the first year. Layers. Cuttings of green shoots in summer in a cool frame. Cuttings of ripe wood, as recommended for retinospora. The named varieties are often grafted on potted common stocks in winter or early fall.

Thyme (*Thymus vulgaris*). *Labiatæ.*

Seeds and division.

Thyrsacanthus (Thyrse Flower). *Acanthaceæ.*

Seeds; cuttings made in spring, and placed in a close, warm frame.

Tiarella. *Saxifragaceæ.*

Seeds and division.

Tigridia (Tiger Flower). *Iridaceæ.*

May be increased by seeds, but generally by offsets.

Tilia (Basswood, Linden, Lime-tree). *Tiliaceæ*.

Stocks are grown from stratified seeds. Layers may be made, and cuttings may be employed, but the named sorts are usually grafted on strong common stocks. Mound-layering is sometimes practiced.

Tillandsia. *Bromeliaceæ.*

May be increased by seeds, and by suckers which should be allowed to grow large before being detached from the parent, and should then be inserted singly in pots, in a compost of loam, peat, and leaf-mold. Keep moderately moist and well shaded. *T. usneoides* is the "Spanish Moss" of the south; rarely propagated, but may be grown from seeds or division of the moss.

Tobacco. See Nicotiana.

Tomato (*Lycopersicum esculentum*). *Solanaceæ.*

Seeds, usually started under glass. Cuttings of growing shoots, rooted under glass, like fuchsias.

Torenia. *Scrophulariaceæ.*

Seeds. Cuttings, in a warm frame.

Torreya. *Coniferæ.*

Increased the same as Thuya and Retinospora.

Trachelium (Throatwort). *Campanulaceæ.*

Seeds and cuttings.

Trachelospermum, Rhynchospermum. *Apocynaceæ.*

Seeds. Firm cuttings, in a frame.

Trachycarpus. *Palmaceæ.*

Seeds and suckers.

Tradescantia. *Commelinaceæ.*

Usually by cuttings; also by seeds and division. See Zebrina.

Trailing Arbutus. See Epigæa.

Trapa (Water Caltrops). *Onagraceæ.*

Seeds.

Tree of Heaven. See Ailanthus.

Trichopilia. *Orchidaceæ.*

Division of the plants. See also under Orchids.

Trichosanthes (Snake Gourd). *Cucurbitaceæ.*

Seeds, either indoors or out.

Tricyrtis. *Liliaceæ.*

Seeds rarely. Offsets and division.

Trillium (Birthwort, Wake-Robin). *Liliaceæ.*

Propagated by seeds and by division.

Triteleia (Triplet Lily). *Liliaceæ.*

Propagated by seeds and by offsets.

Tritoma. See Kniphofia.

Tritonia, including Montbretia. *Iridaceæ.*

Young plants are raised from seeds; but generally increased by division.

Trollius (Globe Flower, Globe Ranunculus). *Ranunculaceæ.*

Seeds. Divisions in early autumn or spring. The seeds should be sown fresh, or a long time will be required for germination.

Tropæolum (Nasturtium, Canary-bird Flower). *Geraniaceæ.*

Seeds, started indoors or in the garden. Tuberiferous species by tubers or division of roots. Perennials sometimes by cuttings in a frame.

Trumpet Creeper. See Tecoma.

Tuberose. See Polianthes.

Tulipa (Tulip). *Liliaceæ.*

Seeds may be sown in boxes of light sandy soil, in late winter, and placed in a coldframe. The next season the young bulbs should be planted in a prepared bed outside. Bulbels may be detached from established bulbs when they are lifted, and grown by themselves. This is the usual method.

Tulip-tree. See Liriodendron.

Tupelo-tree. See Nyssa.

Turnip (*Brassica*). *Cruciferæ.*

Seeds, where the plants are to remain.

Tydæa. As for Gesnera.

Typha (Bullrush, Cat-Tail, Reed Mace). *Typhaceæ.*

Propagation may be effected by seeds sown in a pot plunged in water nearly to the level of the soil; or by division.

Ulex (Furze, Gorse, Whin). *Leguminosæ.*

Propagated by seeds or by cuttings.

Ulmus (Elm). *Urticaceæ.*

Usually propagated by seeds. The seeds of most elms germinate the year they mature (they ripen in spring), and they may be sown at once. The slippery elm (*U. fulva*), however, generally germinates the following year, and the seeds should be stratified. Layers are sometimes made, and suckers may be taken. The varieties are grafted on common stocks.

Umbrella-plant. See Cyperus.

Unicorn Plant. See Martynia.

Uvularia, including Oakesia (Bellwort). *Liliaceæ.*

Seeds; usually by division.

Vaccinium (Swamp Huckleberry, Whortleberry, Blueberry, Billberry, Cranberry). *Ericaceæ.*

Seeds, layers, root cuttings, and divisions of the old plants. Some species by hard-wood cuttings, for which see Cranberry. Huckleberry seeds are small and somewhat difficult to grow. The seeds should be washed from the fruits and stored in sand in a cool place until late in winter. They are then sown in pans or flats on the surface of a soil made of equal parts sand and loam. Cover with fine sphagnum and keep in a cool house or frame, always keeping the seeds moist. Seeds treated in this way may be expected to germinate in a month or two, although they may lie dormant a year. Transplant frequently and keep shaded until large enough to shift for themselves. Layers should be tongued. Cuttings, 2 or 3 inches long, of the best roots, made in fall and placed in mild bottom heat in early spring, often give fair satisfaction. Native plants can be obtained from the woods and fields which will give good satisfaction if small specimens are taken. Gaylussacias are handled in the same way as vacciniums.

Valeriana (Valerian). *Valerianaceæ.*

Seeds and divisions.

Vallota. *Amaryllidaceæ.*

Bulbels, which usually appear above the surface of the pot. Division of the bulbs.

Vanda *Orchidaceæ.*

The majority are propagated in the same way as described for aërides, but two species—*V. teres* and *V. Hookeri*—both tall and quick growing, may be cut into lengths of a few inches. The practice of the most successful cultivators is to start them every year as cuttings about a foot long. See also under Orchids.

Vanilla. *Orchidaceæ.*

Division and cuttings. The vanilla of commerce (*V. planifolia*) is propagated from cuttings, which are planted at the base of trees, upon which the plant climbs. See also under Orchids.

Vegetable Oyster. See Salsify.

Veitchia. *Palmaceæ.*

Seeds.

Veratrum (False or White Hellebore). *Liliaceæ.*

Young plants are obtained by seeds or by division.

Verbascum (Mullein). *Scrophulariaceæ.*

All are raised from seeds sown in any ordinary soil, except *V. nigrum* and *V. pinnatifidum*, which should be increased by division and cuttings respectively.

Verbena (Vervain). *Verbenaceæ.*

Seeds; also by cuttings of vigorous shoots. Some species by division.

Vernonia (Ironweed). *Compositæ.*

May be raised from seeds, division or cuttings, depending on the character of the plant.

Veronica (Speedwell). *Scrophulariaceæ.*

Seeds and division. Shrubby sorts often by cuttings.

Vesicaria (Bladder-pod). *Cruciferæ.*

Annuals by seeds; perennials by division.

Vetch (*Vicia sativa*, etc.). *Leguminosæ.*

By seeds, in open air.

Viburnum. *Caprifoliaceæ.*

Seeds, which should be stratified. They usually remain dormant the first year. Layers usually make the best plants. Green cuttings made in summer and handled in frames give excellent results. *V. tomentosum* (*V. plicatum* of nurseries) is propagated by cuttings. Ripe cuttings are sometimes used for the soft-wooded species. The snowball or guelder-rose (*V. Opulus*) is rapidly increased by layers. It is also a good stock for closely related species. *V. Lantana* and *V. dentata* are good stocks on which varieties difficult to handle can be worked by the veneer-graft during winter.

Victoria (Royal Water Lily, Water Platter). *Nymphæaceæ.*

The seeds should be kept in vessels of water until time for sowing, when they may be placed in loamy soil, and the pot submerged a couple of inches in water, the temperature of which should not be allowed to fall below °85. The tank should be in a light position near the glass. Annual.

Vigna. *Leguminosæ.*

Seeds. The cow-pea or black-pea (*Vigna Sinensis*) by seeds when danger of frost is past.

Vinca (Periwinkle, Running Myrtle). *Apocynaceæ.*

Increased by seeds, and (chiefly) by division. *V. rosea* annually by seeds.

Viola (Violet, Heartsease, Pansy). *Violaceæ.*

The named violets are increased by cuttings made in a cool house from vigorous shoots. Common species by seeds, runners, and division of the plants. Pansies are usually grown from seeds, but named varieties may be multiplied late in the season from cuttings or from layers.

Virgilia. See Cladrastis.

Virginia Creeper. See Ampelopsis.

Virgin's Bower. See Clematis.

Viscum (Mistletoe). *Loranthaceæ.*

Raised from seed, which should be inserted in a notch cut in the bark or under side of a branch of the host. Avoid crushing the seed, and have the embryo directed towards the trunk. To prevent birds from disturbing the seeds after being placed in position, cover with light-colored cloth. The seed may also be fastened to a smooth part of the tree by the sticky substance surrounding it, but more seed is lost. Our native phoradendron can be handled in the same way.

Vitex (Chaste-tree). *Verbenaceæ.*

Seeds. Suckers. Layers. Cuttings of green or ripened wood.

Vitis. See Grape.

Vochysia, Cucullaria. *Vochysiaceæ.*

Seeds ; by ripened cuttings in sand under glass, in heat.

Volkameria. See Clerodendron.

Waahoo. See Euonymus.

Wahlenbergia. See Platycodon.

Waldsteinia. *Rosaceæ.*

May be multiplied by seeds, or by division.

Wallflower (*Cheiranthus Cheiri*). *Cruciferæ.*

Propagated by seeds ; the plants, however, will not flower at the north until the second season ; protection of a frame is required.

Wallichia, Wrightia. *Palmaceæ.*

May be increased by seeds ; or by suckers, which should be gradually separated so as to allow them to make sufficient roots before they are quite detached.

Walnut. See Juglans.

Wandering Jew. See Zebrina.

Water-Cress (*Nasturtium officinale*). *Cruciferæ.*

Cuttings of the young stems, which root in mud with great readiness. Seeds scattered in the water or mud

Water-lily. See Nymphæa, Nelumbo and Victoria.

Water-melon (*Citrullus vulgaris*). *Cucurbitaceæ.*

Seeds, usually sown where the plants are to remain, after the weather is warm and settled.

Watsonia (Bugle Lily). *Iridaceæ.*

The plants a e multiplied by seeds or by offsets.

Wax-flower. See Hoya.

Weigela. See Diervilla.

Wellingtonia. See Sequoia.

Whin. See Ulex and Cytisus.

White Cedar. See Chamæcyparis and Thuya.

White-wood. See Liriodendron and Tilia.

Whitlavia. See Phacelia.

Whortleberry, Huckleberry (*Gaylussacia resinosa*). *Ericaceæ.*

Propagated by seeds, which should be stratified and otherwise carefully handled. See also Vaccinium.

Willow. See Salix.

Wind-flower. See Anemone.

Windsor, Broad or **Horse Bean** (*Vicia Faba*). *Leguminosæ.*

Propagated by seeds in open air after the soil is fairly warm.

Wineberry (*Rubus phœnicolasius*). *Rosaceæ.*

Increases readily by "tips," the same as the black raspberry; also by root cuttings.

Winter Aconite. See Eranthis.

Winter Cress. See Barbarea.

Wistaria. *Leguminosæ.*

Readily grown from seeds. Sometimes by division. Layers. Cuttings of ripened wood, usually handled under glass. The common purple and white kinds are largely grown from root cuttings, an inch or two long, placed in bottom heat, when they will start in four or five weeks. Many of the fancy kinds, especially when wood is scarce, are root- or crown-grafted upon *W. Sinensis*.

Witch-hazel See Hamamelis.

Woodbine. A name properly belonging to climbing Loniceras, but often applied to Ampelopsis, both of which see.

Wormwood, Southern Wood (*Artemisia Absinthium*). *Compositæ.*

Seeds and division.

Wrightia, Balfouria (Palay, or Ivory-tree). *Apocynaceæ.*

Seeds; usually by cuttings, which root readily in sand in heat.

Xanthoceras. *Sapindaceæ.*

Usually multiplied by seeds; root cuttings are sometimes used.

Xanthorrhiza, Zanthorhiza. *Ranunculaceæ.*

Seeds and suckers.

Xanthorrhœa (Black Boy, Grass-tree). *Juncaceæ.*

Seeds; but usually by offsets.

Xanthosoma, including Acontias. *Aroideæ.*

May be increased by cutting up the stem or rootstock into small pieces and planting these in light soil, or cocoa fiber, in bottom heat. After a stem has been cut off a number of shoots are developed, which can be treated as cuttings.

Xerophyllum. *Liliaceæ.*

May be propagated by seeds and by division.

Xiphion. See Iris.

Xylophylla. See Phyllanthus.

Yam. See Dioscorea.

Yellow-wood. See Cladrastis.

Yew. See Taxus.

Yucca (Adam's Needle, Bear's Grass, Spanish Bayonet). *Liliaceæ.*

Increased by seeds; and by divisions, which may be planted in the open ground, or by pieces of thick, fleshy roots, cut into lengths, and inserted in sandy soil, in heat.

Yulan. See Magnolia.

Zamia. *Cycadaceæ.*

Division of the crowns when possible; or by seeds and suckers. The plants are oftenest imported directly from the tropics. See Cycas.

Zanthorhiza. See Xanthorrhiza.

Zanthoxylum (Prickly Ash). *Rutaceæ.*

Seeds, suckers, but more often by root cuttings.

Zea. See Maize.

Zebra-grass. See Miscanthus.

Zebrina (Wandering Jew). *Commelinaceæ.*

Very easily multiplied by single-joint cuttings of the trailing shoots. These shoots root at the joints if allowed to run on moist earth.

Zephyranthes, including Habranthus (Flower of the West Wind, Zephyr Flower). *Amaryllidaceæ.*

May be multiplied by seeds; or by separating the bulbels.

Zingiber, including Zerumbet (Ginger). *Scitamineæ.*

Propagated by division.

Zinnia (Youth-and-Old-Age). *Compositæ.*

Seeds, sown either indoors or out.

Zizania (Wild or Indian Rice). *Gramineæ.*

Seeds, sown along water courses or in bogs in fall or spring.

Zizyphus. See Jujube.

Zygadenus, including Amianthemum. *Liliaceæ.*

Readily multiplied by seeds or by division.

Zygopetalum. *Orchidaceæ.*

Division. See also under Orchids.

Zygophyllum (Bean Caper). *Zygophyllaceæ.*

Seeds, when they can be had; otherwise by cuttings in a frame.

GLOSSARY.

Adventitious. Said of buds which appear in unusual places, especially of those which are caused to appear on roots and stems by any method of pruning or other treatment.

Air-layering. See Pot-layering.

Bark-grafting. A kind of grafting in which the cions are inserted between the bark and wood of a stub; often, but erroneously, called crown-grafting. Fig. 135.

Bottle-grafting. A method of grafting in which a shred of bark from the cion, or a portion of the bandage, is allowed to hang in water (generally in a bottle, whence the name), for the purpose of supplying the cion with moisture until it has united with the stock. Pages 112, 132.

Bottom heat. A term used to designate the condition that arises when the roots of plants, or the soil in which they grow, are exposed to a higher temperature than that of the air in which the aërial portions of the same plants are growing.

Breaking. Said of buds and cions which start (or *break*) prematurely. Page 101.

Brick (of mushroom). See Spawn.

Bud. As used by propagators, a *bud* comprises the leaf-bud (or rarely a short spur) and a bit of bark or wood to which it is attached. Figs. 85, 96, 99, 100.

Budding. The operation of applying a single bud to the surface of the growing wood of the stock, with the intention that it shall grow. The bud is usually inserted underneath the bark of the cion, and is held in place by a bandage. Budding is a part of the general process of graftage. Called *inoculation* in old writings. Page 94.

Bulb. A large and more or less permanent and fleshy leaf-bud, usually occupying the base of the stem and bearing roots on its lower portion. Scaly bulbs, like that of the lily (Figs. 19, 20), are made up of narrow and mostly loose imbricated scales. Tunicated or laminated bulbs, like that of the onion, are composed of closely fitting and more continuous layers or plates.

Bulbel. A smaller or secondary bulb borne about a mother bulb ; bulbule. Page 27.

Bulblet. A small bulb borne wholly above ground, usually in the inflorescence or in the axil of the leaf, as in "top onions," tiger lily, etc.

Bulbo-tuber. See Corm.

Bulbule. See Bulbel.

Callus. The mass of reparative or healing tissue which forms over a wound.

Cambium. The tissue which lies between the bark and the wood, and from which those parts arise.

Chinese-layering. See Pot-layering.

Chip-budding. That style of budding which removes a truncheon or chip of bark and wood from the side of the stock, and fills the cavity with a similarly shaped bud from the variety which it is desired to propagate. Fig. 100.

Cion or Scion. A portion of a plant which is mechanically inserted upon the same or another plant (stock), with the intention that it shall grow. See Chapter V., and Figs. 101, 102, 108, 111, 115, 116, 119, and others.

Cion-budding. See Shield-grafting.

Circumposition. See Pot-layering.

Cleft-grafting. That method of grafting in which the stock is cut off completely and then split, and one or more cions, cut wedge-shape, inserted in the cleft. Figs. 118, 119, 124, 125, 126, 134.

The grafted end of the stock (Figs. 118, 126) is called a *stub*.

Clove. One of the small separable portions of a composite bulb, as in the garlic.

Corm. A solid bulb-like tuber, more or less covered with a sheathing or enwrapping tissue, as in the crocus and gladiolus; bulbo-tuber. Fig. 25.

Cormel. A small corm borne about another or mother corm. Page 31. Fig. 25. Sometimes called *spawn*.

Crown. A detachable portion or branch of a rootstock bearing roots and a prominent bud. Page 32.

That portion of the plant at the surface of the ground which stands between the visible stem and root; collar.

Crown-grafting. Grafting upon the crown or collar of a plant (*i. e.*, at the surface of the ground). Page 107.

Also applied to bark-grafting (but improperly). Page 129.

Cuttage. The practice or process of multiplying plants by means of cuttings, or the state or condition of being thus propagated. [First used by the present author in 26th Report of the State Board of Agriculture of Michigan, p. 432 (1887). Equivalent to the French *bouturage*.] See Chapter IV.

Cutting. A severed portion of a plant, inserted in soil, water, or other medium, with the intention that it shall grow; a slip. See Chapter IV.

Cutting-bench. A bed or table in a glass-house, or under cover, used for the rooting of cuttings. Fig. 58.

Cutting-grafting. The grafting of a cion upon a cutting. Page 131. Figs. 105, 137.

Damping-off. The rotting away of the tissue of plant stems at or near the surface of the ground. It is the work of fungi or of germs. Page 23.

Division. Propagation of plants by means of severed portions of the root system or of subterranean stems. Pages 32 and 58.

Double-working. Grafting or budding upon a plant or shoot which is itself a graft or bud. Page 133. Fig. 140.

Dressing (of stocks). The trimming of stocks, which are to be budded or grafted, before they are set in the nursery row. Fig. 86. Page 96.

Dwarfing. The permanent checking of the growth of a plant so that it never attains its normal stature. The chief means employed in the dwarfing of trees are, grafting upon a slow-growing stock, heading-in of the top, confining or pruning the roots. Page 147.

Eye. A bud. Single-eye cuttings are those bearing but one bud. Fig 66.

Flagging. Wilting. Said of plants newly transplanted, or of cuttings. Pages 53, 66, 68. Also applied to the general wilting of plants due to lack of water.

Flat. A shallow box or tray, in which the gardener grows or transports plants.

Flute-budding. That method of budding which removes a rectangular portion of bark from the stock and fills the cavity with a similar piece of bark, of the desired variety, bearing a bud. Fig. 99.

Frame. The structure forming the sides and ends of cold-frames or hotbeds. A frame is commonly six feet wide and of sufficient length to accommodate from three to six three-feet-wide sash. It is usually made of boards. The area covered by a single glass shutter is called a *sash*, and is generally 3 x 6 feet. The area which is commonly understood as one *frame* is 6 x 12 feet.

Free Stock. See Stock and Seedling

Germination. The act or process by means of which a seed or spore gives rise to a new plant. Germination is complete when the plantlet has exhausted the store of food in the seed and is able to support itself. Page 9. The word germination cannot be properly applied to the arising of plants from tubers, as of the potato ; ***vegetation*** is the better term in such cases.

Graftage. The process or operation of grafting or budding, or the state or condition of being grafted or budded. [First used by the present author in 26th Report of the State Board of Agriculture of Michigan, p. 433 (1887). Equivalent to the French *greffage*.] See Chapter V.

Grafting. The operation of inserting a cion in a stock. It is commonly restricted to the operation of inserting cions of two or more buds, in distinction from budding, or the operation of inserting a single bud in the stock; but there are no essential differences between the two operations. See Chapter V.

H-budding. Much like flute-budding (which see), except that the bark which is loosened from the stock is left attached in two flaps, secured at the upper and lower ends, and these flaps are tied over the bud. Fig. 98.

Heading-in. Cutting back or shortening the shoots or branches of plants, in distinction to removing the branch bodily at its point of union with the parent branch.

Heel. A form of cutting of which the lower end comprises the very base of the shoot as it grew upon the parent branch. Fig. 60.

Heeling-in. The temporary covering of plants, or of their roots, in order to preserve or protect them until they are placed in permanent quarters.

Herb. A plant which dies to the ground once a year, at the approach of winter or of the inactive season. Used in distinction to woody plants, like shrubs and trees. Perennial herbs are those of which the tops or aërial portions perish while the root lives on from year to year, in distinction to an annual herb, which perishes outright after one season of growth and flowering.

Herbaceous-grafting. The grafting of soft, growing shoots; generally confined to the grafting of herbs, but the term may be applied to the grafting of the growing shoots of woody plants. Page 130.

Inarching. The joining, by graftage, of parts of two contiguous plants, whilst the cion part is still attached to its parent plant. The cion is severed from its parent when it has united with its foster stock. Figs. 138, 139.

Inlaying. The insertion of a cion into a cavity or notch made by the removal of a piece of wood from the stock. Fig. 116.

June-budding. The practice of budding trees very early in the season, commonly in June, with the expectation that the buds are to grow the same season. Practiced in the south. Page 103.

Knaur. An excrescence or burr appearing as redundant or adventitious tissue upon a woody plant, and which may be used for the propagation of the plant when removed and treated like a cutting. Page 64.

Layer. A shoot or root, attached to the parent plant, partially or wholly covered with earth, with the intention that it shall take root and then be severed from the parent. See Figs. 29, 30.

Layerage. The operation or practice of making a layer, or the state or condition of being layered. [Word first used by the present author in 26th Report of the State Board of Agriculture of Michigan, p. 431 (1887). Equivalent to the French *marcottage*.] See Chapter III.

Mallet. A form of cutting in which a prominent transverse portion of an older branch is left upon the lower end. Fig. 61.

Mother-bulb. The large or parent bulb about which smaller bulbs, or bulbels, are borne.

Mound-layering. The rooting of upright shoots by means of heaping earth about them; stool-layering. Fig. 32.

Nursery. An establishment for the rearing of plants. In America the word is commonly but erroneously used in connection with the propagation of woody plants only, as fruit trees and ornamental trees and shrubs. The word properly includes the propagation of all plants by whatever means, and in this sense it is used in this book.

Offset. A rosette or cluster of leaves, on a very short axis, borne next the surface of the ground, and in time becoming detached and making an independent plant. Page 32.

Piece-root-grafting. Grafting upon pieces of roots. Page 109. Figs. 103, 104. See also, Figs. 144–152.

Pip. A perpendicular rootstock or crown, used for the propagation of a plant. Page 33.

A seed of one of the "seed fruits," as apple or pear. Used in distinction to the stone or pit of a "stone fruit."

Plate-budding. Much like flute-budding (which see), except that the bark loosened from the stock is left attached at its lower end, and this flap is then raised and bound over the bud. Fig. 97.

Pot-layering. The rooting of an aërial stem by means of encircling it with earth or moss. Figs. 34, 35. Known also as air-layering, Chinese-layering, and circumposition.

Prong-budding. A method of propagation much like shield-budding, except that the bud bears a short branch or spur. Fig. 96.

Pseudo-bulb. Generically, a corm, or homogeneous bulb-like enlargement under ground or at the base of the plant. The term is now commonly restricted, however, to the thickened bases of the stems in various orchids; these usually stand just at or below the surface of the ground. See under Orchids, page 284.

Regermination. The continuation or resumption of the process of germination after it has been completely interrupted or checked. Page 9.

Rhizome. A subterranean branch or stem; rootstock. A rhizome is distinguished from a root by the presence of joints and buds; and it is usually thicker and more fleshy than the roots.

Root-grafting. Grafting upon a root. Pages 107, 109. Figs. 103, 104. See also, Figs. 144–152.

Rootstock. See Rhizome.

Root-tip. See Tip.

Saddle-grafting. That method of grafting in which the stock is cut wedge-shape, and the cion is cleft and slipped down over the wedge. Fig. 109.

Scion. See Cion.

Seed. The reproductive body which results from a flower, and which is the product of sexual union. It is a ripened ovule. The rudimentary plantlet which it contains is the *embryo*.

Seedage. The process or operation of propagating by seeds or spores, or the state or condition of being propagated by seeds or spores. [Word first used by the present author in 26th Report of the State Board of Agriculture of Michigan, p. 430 (1887).] See Chapter I.

Seed-grafting. The insertion of a seed, as a cion, in a stock. Page 131.

Seedling. A plant growing directly from the seed without the intervention of graftage. If it is used upon which to graft or bud, it is known as a *free stock*.

Separation. The act or process of multiplying plants by means of naturally detachable vegetative organs, or the state or condition of being so multiplied. [First technically used in this meaning in 26th Report of the State Board of Agriculture of Michigan, p. 432 (1887).] See Chapter II.

Set. An indefinite term applied to various vegetative parts which are used for purposes of propagation. It may designate a root cutting (Fig. 64), or a small bulb (as in the onion).

Shield-budding. That method of budding which makes a T-shaped incision on the stock (Fig. 90), and inserts a shield-shaped bud (Fig. 85) into the opening (Fig. 91). Page 95.

Shie'd-grafting. The insertion of a cion with a wedge-shape point into an incision like that used for shield-budding; cion-budding. Fig. 115.

Side-grafting. The insertion of a cion with a sharp or wedge-shape point into a diagonal incision into the wood on the side of the stock. Figs. 113, 114, 115.

See, also, Veneer-grafting.

Silver Sand. Clear white sand devoid of organic matter, used for the starting of cuttings. Page 54.

Slip. See Cutting; also, page 65.

Spawn. The dried mycelium of mushrooms, preserved fn dense masses of prepared earth and manure (known as "bricks"), or in loose, strawy litter (known as "French spawn"), or in the loose earth of mushroom beds.

See, also, Cormel.

Splice-grafting. The joining of simple oblique surfaces in the stock and cion. Fig. 110.

Spore. The reproductive body of a flowerless plant (or cryptogam), as of ferns, fungi, sea-weeds, and the like. It has no embryo, and it commonly consists of a single cell.

Spur. A very short branch, usually lateral, which does not increase markedly in length from year to year. The normal office of spurs is to bear flowers and fruit.

Stem-grafting. Grafting upon the stem or trunk of a plant, between the crown (or the ground) and the top. Page 107.

Stick. A twig of the recent growth of any plant, bearing buds which it is proposed to use in propagation. Fig. 87.

Stock. In graftage, a plant or part of a plant upon which a cion or bud is set. A *free stock* is a seedling, in distinction from a grafted stock.

Stolon. A decumbent shoot which, without artificial aid, takes root and forms an independent plant. The honeysuckles, some osiers (as *Cornus stolonifera*), and many other bushes with long and slender branches, propagate by means of stolons. The black raspberry propagates by a special kind of stolon, rooting only at its tip.

Stool. A clump or mass of roots or rootstocks which may be readily divided for purposes of propagation. Fig. 27.
An established root of a bush-like plant from which shoots are grown for the purpose of layering. Page 39.

Stool-layering. See Mound-layering.

Stove. The warmest portion or compartment of glass-houses, used for tropical plants.

Stratification. The operation of burying seeds, usually in layers, in order to keep them viable and to soften their integuments. Page 15.

Striking. A term applied to the forming or emitting of roots on layers or cuttings. A cutting is *struck* when it has made roots.

Stub. See Cleft-grafting.

Sucker. A shoot sent up from the roots, more particularly one which it is not desired shall grow. Also used for strong and mischievous shoots in the top of a tree. See Watersprout.

Tip. The plant formed at the end of a stolon when it strikes the ground. The black raspberry propagates naturally by tips. Fig. 28. The branches of other plants, like the currant, can be made to strike at the tip when they are bent over and fastened in the ground.

Tongue-grafting. See Whip-grafting.

Top-grafting. Grafting the top or branches of a tree or bush. Pages 107, 122. Figs. 127–132.

Tuber. As used in this book, a prominently thickened and turgid homogeneous portion of a root or stem, usually subterranean, and which generally does not increase or perpetuate itself (as bulbs and corms do) by direct offshoots or accessions. Page 32.

In botanical writings, the term is commonly restricted to thickened subterranean stems, as in the Irish potato and the Jerusalem artichoke.

Tubular-budding. See Whistle-budding.

Veneer-grafting. That style of grafting in which a cion, with the bark removed from one side, is applied to the side of a stock from which a strip of bark has been removed. Sometimes called side-grafting. Figs. 111, 112.

Watersprout. A strong and comparatively soft shoot arising from an adventitious bud in the top or from the trunk of a plant. It is usually forced out by severe pruning or heading-in. It is an undesirable type of shoot when fruits or flowers are desired, because it expends its energies for one or several seasons in exuberant growth. Watersprouts are often purposely obtained, however, when it is desired to secure young wood in which to set buds in old trees. Page 105. Compare Sucker.

Whip-grafting. A style of grafting in which the stock and cion are shaped alike—an oblique cut and a perpendicular cleft; tongue-grafting. Figs. 101, 102.

Whistle-budding. That kind of budding in which a ring or girdle of bark is removed from the stock, the girdle being filled by a similar ring, with a bud attached, of the variety which it is desired to propagate; tubular-budding. Page 106.

INDEX.

The alphabetical entries or headings in the Nursery List, Chapter VI., are not included in this index, but all secondary and incidental names and references comprised in it are intended to be included here. Inasmuch as many cultivated plants of secondary importance had to be omitted from the Nursery List in the effort to economize space, the present index has been made to include the names of all the natural families of plants which that List comprises, in the hope of aiding the student in apprehending the general methods of propagation which apply to the family of which any plant, itself omitted from the List, may be a member. This ordinal index was made a separate feature of the first edition of the book.

Page

Acanthaceæ 159, 169, 183, 231, 236, 260, 265, 323, 327
Achras 315
Aconite 160
—Winter 231
Acontias 335
Acorns, transporting 19
Adamsia 306
Adonis seeds 8
Ægle trifoliata 76
Affinities, for graftage 77
African Lily 162
Agrostemma 268
Air-layering 40
Akebia quinata 68
Alcoholic waxes 136
Alismaceæ 163, 193, 266
Allegheny Vine 161
Alligator Pear 295

Page

Allium Ascalonicum 318
—Cepa 281
—fistulosum 281
—Porrum 264
—sativum 233
—Schœnoprasum 209
—Scorodoprasum 311
Allspice 193
Almond for peach stock . . . 74
—Tropical 326
Althea cuttings 68
Amarantaceæ 165, 258
Amaryllidaceæ . 162, 165, 188, 190, 192, 196, 197, 213, 218, 220, 222, 228, 232, 233, 238, 239, 249, 253, 256, 257, 258, 265, 277, 279, 289, 303, 331, 336
Amelanchier oblongifolia . . 260
American Agriculturist, quoted 246

Page
American Centaury 313
—Cress 182
Amomophyllum 320
Anacardiaceæ 167, 271, 300, 311, 316, 321
Ananas sativus 298
Andromeda seeds 20
Anethum graveolens 226
Aniseed-tree 257
Anisopetalum 193
Annular-budding 106
Anonaceæ 168, 176, 178
Apium graveolens 203
—Petroselinum 289
Apocynaceæ 164, 166, 170, 227, 270, 279, 325, 328, 332, 334
Apple Berry 186
—dwarfing 148
—effect on land 140
—Paradise 39
—root cuttings 61
—seeds, treatment of 17, 19
—stock, height for 146
—stocks for 74, 75
Apricot, St. Domingo 270
—stocks 164
Aquatic plants, sowing seeds . 21
Aquilegias 33
Araliaceæ 175, 221, 240, 250, 286, 316
Arbor-vitæ, Japanese 309
Arching layers 38
Aristolochiaceæ 176, 177
Arloing, on Cactus cuttings . . 194
Aroideæ 160, 164, 166, 169, 176, 177, 195, 249, 274, 287, 294, 297, 311, 315, 316, 320, 322, 335
Artemisia Absinthium 334
" Dracunculus . . . 326
Arthur, J. C., Geneva tester 11
Artillery Plant 298
Asclepiadaceæ 167, 177, 205, 241, 255, 271, 295, 321, 322
Asexual propagation 91

Page
Ash 65
Aspen 304
Asphodel 178
Atkinson, quoted 23
Aubletia 288
Augur, Mr., quoted 75
Autumn Crocus 214
Avocado Pear 295
Azalea Indica 65
Balfouria 334
Balsam Fir stocks . . . 157, 298
Baltet, quoted 80
Bamboo 181
Bandages, waxed 137
Bark-grafting 129
Barnard's tank 49
Bartow Courier-Informant, quoted 283
Bass-bark 100
Bastard Cedar 203
—Indigo 166
Basswood cuttings 55
Bead-tree 273
Bean, Broad or Horse . . . 334
—grafting 78
—dwarfs 148
Bearbind 198
Beard-tongue 294
Bear's Breech 159
Beech, grafting 116
Begonia 70, 71, 72
—plantlets of 30
—seeds 21
Begoniaceæ 184
Belladonna 179
Bell-jar 44
Bellwort, Giant 287
Bene 318
Berberidaceæ 163, 182, 186, 203, 230, 259, 264, 303
Berberis vulgaris 182
Beta vulgaris 184
Bignoniaceæ 166, 186, 202, 209, 229, 258, 326
Bindweed 198, 215

Page

Bird of Paradise Flower . . . 323
Bird's-tongue Flower 323
Blackberry Lily 185
—root cuttings 61
Black Boy 335
—Hellebore 252
—Pea 332
—Salsify 316
Bladder-pod 331
Blue Cohosh 203
Blue Flag 258
Boards, on seeds 4
Boneset 233
Boning 98
Boring seeds 18
Borraginaceæ 167, 176, 189, 215, 251, 273, 276, 281, 282, 325
Botanical Gazette, quoted 11
Bottle-grafting 112, 132
Bottom heat for cuttings 53
— —for seeds 8
Botryanthus 275
Bouvardia cuttings 61
Bowstring Hemp 315
Boxberry 238
Bradlev, quoted 77
Brake, Bracken 306
Bramble 313
Brassica oleracea 194, 261
—species 276, 329
Broad Bean 334
Broad-leaved China Fir . 220
Bromeliaceæ . . 186, 191, 298, 328
Broom, Scotch 222
Brush screen 5
Bryophyllum 70
Buckthorn, Sea 253
Budded trees 148
Budding 94
Budd, Professor, quoted . . . 74
Bud-grafting 79
Bugle 163
—Lily 333
Bulbels 27, 60
Bulblet 30, 60

Page

Bulbs 26
Burbidge, F. W., quoted . . 84, 92
Burseraceæ 181, 189, 193
Butterfly 316
Butternut 201, 259
Bitter Vetch 287
Button-wood 204
—Snake-root 265
—tree 215
Cabbage, leaf cuttings . . . 70
Cactaceæ 194, 231, 282, 294, 295, 297, 310
Cactus cuttings 60, 65, 67
—grafting 128
Cælestina 163
Calampelis 229
Calceolaria, sowing 20
Calico-bush 261
California Poppy 232
Calipers 143
Callistemma 197
Callus 55
Calycanthaceæ 198, 209
Caltrops, Water 328
Cambium 78
Cambogia 238
Camellia cuttings 65
—graftage 76
Campanulaceæ . . . 161, 199, 221, 222, 287, 298, 300, 308, 328
Campanula Rapunculus . . . 308
Camphor-tree 199
Canary-bird Flower 329
Candleberry 277
Canellaceæ 199
Canna 33, 34
—treatment of seeds 18
Cape Primrose 323
Caper 200
Capparidaceæ 200, 212
Caprifoliaceæ 157, 226, 266, 268, 314, 325, 332
Caprifolium 268
Carnation cuttings 65, 66
—layering 38

Page
Carrion Flower 321
Carrot, seed tests 10
Carthamus tinctorius 314
Caryophyllaceæ 175, 200, 249, 268, 299, 315, 318, 325
Castalia 280
Castor Bean 311
Casuarineæ 202
Cayenne 295
Cedar, Bastard 203
—Japan 219
Celastraceæ 202, 203, 276
Celery, seed tests 10
—sowing 4, 22
Cellars 143
Centaury 232
—American 313
Cercis Japonica 68
Cereus, species 194
Chærophyllum bulbosum . . . 208
Chaste-tree 333
Chauvière's oven 47
Cheiranthus Cheiri 333
Chenopodiaceæ . 184, 189, 282, 321
Cherry, dwarfing 147
—effect on land 140
—root cuttings 61
—stock, height for . . . 146, 147
China Root 319
Chinese layering 40
—Sacred Lily 277
Chip-budding 107
Chisels 119
Chocolate-tree 327
Choko 317
Christmas Rose 252
Chrysanthemum 66
—fœniculaceum 271
—frutescens 271
—grafting 78
Cichorium Endivia 230
—Intybus 208
Cineraria, sowing 20
Cion-budding 116
—grafting 79

Page
Cion-cutting 107
Circumposition 40
Cistaceæ 211, 251
Citrullus vulgaris 333
Citrus fruits, stocks for 76
—Aurantium 282
—Decumana 304
—Japonica 262
—Limetta 266
—Medica 212
— —var. acris 266
— —var. Limon 264
—species 283
—trifoliata 76, 161
Cives 209
Classification of graftage . . . 79
Cleft-grafting 118
Climate and graftage 75
Clintonia 213, 228
Cloche 44
Cloth screens 6
Club-moss 269
Cob-nut 217
Cocoanut fiber for seeds . . . 20
Coco Plum 210
Cocos nucifera 214
Coffee-tree 214
Cohosh, Blue 203
Coleuses 65, 66
Collodion for wounds 138
Color modified by graftage . . 76
Compass-plant 318
Combretaceæ 215, 326
Comfrey 62
Commelinaceæ . . 215, 226, 328, 336
Compositæ . . . 159, 162, 163, 166, 169, 176, 177, 178, 185, 188, 190, 196, 197, 201, 202, 208, 209, 210, 211, 216, 217, 222, 223, 228, 229, 230, 233, 234, 238, 239, 248, 249, 251, 252, 257, 263, 264, 265, 271, 276, 287, 299, 313, 314, 316, 317, 318, 320, 322, 325, 326, 331, 334, 336
Cone Head 323
—flower 313

Page

Coniferæ 157, 175, 197, 203, 205, 219, 220, 222, 236, 240, 252, 260, 262, 297, 298, 299, 303, 306, 309, 316, 317, 326, 327, 328
Conifer cuttings 57, 64
—grafting 115
Convolvulaceæ 176, 198, 215, 257, 279, 324
Coquito Palm 259
Coral-tree 232
Coriandrum sativum 216
Cork for seeds 20
—tree 296
Cormels 31
Corms 31
Cornaceæ . . 179, 185, 216, 238, 280
Cornell Exp. Sta., quoted . . 9, 23, 111, 117, 139, 148, 174
Corn, fertility in 140
Corypha 205
Cottonwood 304
Cow-dung for seeds 20
Cow-itch 275
Cow Parsley 252
—Parsnip 252
Cowslip, American 228
Crabs as stocks 76, 170
Crakeberry 230
—Portugal 216
Crambe maritima 317
Crandall, C. S., quoted 88
Crape Myrtle 262
Crassulaceæ . 192, 217, 218, 261, 317
Cress, American or Upland . . 182
--Curled 294
Crocus 31
—autumn 214
—sativus 314
Crosswort 219
Crowberry 216, 230
Crown-grafting 107, 129
Crowns 32, 33
Cruciferæ 165, 167, 174, 179, 182, 194, 200, 218, 225, 228, 232, 252, 254, 256, 261, 268, 270, 272, 294, 308, 311, 317, 329, 331, 333

Page

Cryptomeria Japonica 64
Cubeb 299
Cucullaria 333
Cucumis Anguria 239
—Melo 273
—sativus 219
Cucurbitaceæ 158, 192, 219, 239, 242, 268, 272, 273, 274, 306, 317, 321, 329, 333
Cucurbita Pepo 242
--species 306, 321
Cupuliferæ 164, 186, 201, 208, 209, 233, 287, 307
Curled Cress 294
Currant cuttings . . . 55, 56, 63
Currants, by tips 36
Cuttage 44
Cutting-bench 52
—grafting 131
Cuttings 44–72
Cutting side-graft 116
Cuttings as stocks 110
Cycad truncheons 65
Cycadaceæ . . 201, 221, 227, 230, 335
Cynara Scolymus 176
Cyperaceæ 200, 221, 289
Cyrillaceæ 222
Cyrtopodium 296
Dahlia 32
—grafting 129
Dahlias, dwarf 148
Dame's Violet 252
Damping-off 23, 54
Darwin, quoted 77, 87, 91
Date Plum 227
Datiscaceæ 224
Delphinium seeds 8
Dendrobium Phalænopsis . . 286
Depth to sow seeds 21
Desert Willow 209
Deutzia cuttings 68
Dewberry 35
Dianthus barbatus 325
Diapensiaceæ 238
Diervilla 67

Page

Dilleniaceæ 199, 227
Dioscoreaceæ 227, 326
Diospyros Kaki 295
—Virginiana 295
Dipsaceæ 316
Dish-cloth Gourd 268
Disocactus 297
Distance apart for trees 146
Dittany 226
Division 32, 58
Dog's-tooth Violet 232
Dorcoceras 181
Double-grafting 133
Double-working 133
Doucin stock 148
Downing, quoted 77
Dracæna 61
Dragon's Head 228
Dressing of stocks 96
Droseraceæ 227, 229
Duck's Foot 303
Dwarfing 73, 74, 147
Dwarf Pears 62
Ebenaceæ 227, 295
Ebony, Mountain 183
Elæagnaceæ 229, 253, 318
Elichrysum 251
Elm 65
Elephant's Foot 326
Empetraceæ 204, 216, 230
Endive, seed tests 10
Ensilage Corn 140
Entomosporium maculatum . 291
Epacridaceæ 167, 179, 230
Equestrian Star 253
Ericaceæ 167, 175, 180, 184, 197, 202, 213, 218, 230, 231, 238, 261, 263, 264, 265, 267, 269, 287, 306, 310, 330, 334
Eruca sativa 311
Ervums 264
Eryngo 232
Euphorbiaceæ 158, 163, 193, 202, 210, 214, 233, 259, 271, 297, 303, 311, 322

Page

Euphorbia cuttings 65
Evergreen cuttings 57, 64
Everlastings 251
Exotic seeds 19
Fadyenia 238
Fagopyrum esculentum . . . 192
—Tataricum 192
False Solomon's Seal 319
Fan Palm 267
Felicia 162, 234
Fenzlia 239
Fern, Flowering 287
—Hartford 269
Ferns, plantlets of 30
—sowing 24
Fertility of lands 139
Ficoideæ 273
Ficus Carica 236
—cuttings 65
—elastica 41
Field, The, quoted 83
Fig, Marigold 273
Filices 224, 226, 234, 269, 287, 300, 304, 306
Filing seeds 18
Fir, Broad-leaved China . . . 220
First-class trees 143
Five-Finger 305
Flavor modified by graftage . 76
Flax Lily 297
Fleur-de-Lis 236
Flowering Fern 287
—Rush 193
Flower of the West Wind . . . 336
Flute-budding 106
Fœniculum 234
Formation of roots 55
Forsyth's cutting-pot 51
French Mulberry 196
Frenela 197
Fringe Flower 316
Fruit-grafting 131
Fuchsias 65
Fuller's Herb 315
Fumariaceæ 161, 217, 226

Page
Fungus in cellars 145
Funkia 33
Galeopsis 321
Garden, quoted 83, 84
Garden and Forest, quoted . . 74
Gardener's Chronicle, quoted . 29
Garland Flower, Indian . . . 251
Gauging 143
Geneva seed tests 11
Gentian 239
Gentianaceæ . 209, 232, 239, 273, 313
Geraniaceæ 180, 181, 232, 239, 287, 294, 329
Geraniums 65, 66, 67, 70, 72
Germander 326
German seed tester 13
Germinators 18
Gesneraceæ 159, 161, 181, 226, 239, 261, 277, 308, 314, 318, 323
Giant Bellwort 287
Ginep 273
Girdles, repairing 129
Girdling layers 38
Gladiolus 31
Glechoma 278
Globe Flower 329
—Mallow 320
—Ranunculus 329
—Thistle 229
Glory Pea 213
Gloxinia 60, 72
—grafting 129
—seeds 21
Glyptostrobus 326
Goat's Rue 238
Gnetaceæ 230
Golden Bell 236
—Chain 262
—Drop 282
Goldfussia 323
Goldy-Locks 210
Goober 175
Goodenovieæ 316
Gooseberries, by tips 36
—mound layering 39

Page
Gooseberry, Barbadoes 295
—cuttings 57, 63
—Otaheite 210
Gossypium 217
Gourd, Dish-cloth 268
—Snake 329
Grades of trees 142
Graftage 73–156
Graft-hybrid 77
Grafting 107
Gramineæ 163, 177, 181, 190, 191, 195, 249, 270, 274, 296, 320, 324, 336
Grape layering 37
Grapes, cuttings of 53, 55, 57, 63, 64
—grafting 111, 112, 116, 117, 121, 132
Grass of Parnassus 289
Grass-tree 335
Green-Briar 319
Green-wood cuttings 65
Ground Cherry 297
Groundsel 317
Gutta-Percha tree 258
Guttiferæ 198, 213, 238, 270
Habranthus 336
Hæmodoraceæ 249, 315
Halorageæ 249
Hamamelideæ, 192, 237, 250, 266, 289
Hardiness and graftage 75
Hard-wood cuttings 62
Hare's Ear 193
Hartford Fern 269
Hatchet Cactus 294
Haws, treatment of seeds . . . 18
H-budding 106
Healing of wounds 126
Hedge Bindweed 198
—Mustard 232
—Nettle 321
Heeling-in 146
Heel of cutting 55
Helianthus tuberosus 177
Heliotrope 66, 251

Page
Hellebore, Black 252
—White 331
Hemerocallis 33
Hemp, Bowstring 315
Hen and Chickens 32
Herbaceous-grafting 130
Herb of Grace 313
Heron's Bill 232
Hews, A. H. & Co., mentioned . 14
Hibiscus esculentus 280
Hicoria Pecan 293
Hickories, cuttings 58
Hickory-nuts, treatment of . . 17
Higginsia 253
Hoit's grafting device 120
Holly, Japan 287
—Mountain 278
Hollyhock, grafting 129
Holly seeds, treatment of . . . 17
Hoop Withy 311
Honey Plant 255
Honeysuckle 174
Hop-tree 306
Horned Rampion 298
Horse Bean 334
Horse-radish sets 57, 61
Hortensia 256
Hoskins' wax 138
House-leek 32
House plants 65, 67
Husk Tomato 297
Husmann, George, quoted . 246
Hyacinth 255
—cuttings 60
—propagation 28
Hyacinthus candicans 238
Hydrangea 67, 68
Hydrophyllaceæ 278, 296
Hypericaceæ 177, 256
Hyssopus officinalis 256
Ice Plant 273
Ilicineæ 257, 278
Illairea 267
Ipomœa Batatas 324
—grafting 129

Page
Impatiens Balsamina 181
—Sultani 181
Inarching 79, 81, 132
Indian Cup 315
—Currant 325
—Garland Flower 251
Indigo 257
Influence of stock and cion . . 74
Inlaying 117
Inula Helenium 230
Iridaceæ 168, 169, 176, 180, 185, 211, 219, 221, 235, 237, 240, 252, 258, 278, 316, 319, 320, 327, 329, 333
Ironweed 331
Ironwood 287
Ivory-tree 334
Jacobæa 317
Jalapa 274
Janipha 271
Japan Cedar 219
Japanese Arbor-Vitæ 309
—Rose 198
Japan Holly 287
Japonica 198
Jasminanthes 322
Jasmine 259
—Box 296
Jessamine 259
Juglandaceæ . . . 253, 259, 293, 306
Juncaceæ 260, 335
June budding 103
June-struck cuttings 67
Juniper 260
Junipers, cuttings 64
Kalmia seeds 20
Keeping qualities of fruit . . 75
Kidney Vetch 169
Kier's layering-racks 42
Kinds of grafting 80
Knaurs 64
Knight, on sowing 2
Knives 97, 111, 119
Knot-Grass 304
Knot-Weed 304

Page
Labiatæ 163, 181, 183, 195, 203, 214, 228, 254, 256, 263, 264, 278, 294, 295, 297, 313, 314, 315, 320, 321, 326, 327
Labrador Tea 263
Lace Bark 262
Lactuca sativa 264
Ladies' Ear Drop 237
Lady's Mantle 163
—Slipper 222
—Smock 200
Lagenaria 242
Lands, management of 139
Lantanas 70
Larch 262
Lath screen 5
Lauraceæ . . 199, 263, 266, 295, 315
Lavender 263
Lead Plant 166
Leadwort 303
Leaf-blight 291
Leaf Cactus 297
—cuttings 60, 70
—grafting 131
Leather Leaf 202
Lecoq's oven 48
Leguminosæ . . 158, 161, 162, 163, 166, 169, 170, 175, 179, 182, 183, 191, 195, 196, 200, 201, 202, 205, 212, 213, 215, 216, 219, 222, 223, 228, 232, 234, 238, 240, 241, 249, 250, 251, 255, 257, 261, 262, 263, 264, 267, 268, 272, 274, 275, 282, 287, 288, 289, 290, 295, 296, 303, 306, 311, 318, 320, 324, 325, 331, 332, 334
Lemon Verbena 165
Lentibulariaceæ 299
Leopard's Bane 228
Lepidium sativum . . . 218, 294
Lepismium 310
Levisticum officinale 268
Layerage 35
Ligularia 317
Light, and germination 8
Lilac, cuttings 67

Page
Lilacs 33
Liliaceæ 162, 163, 164, 169, 178, 183, 186, 188, 190, 192, 193, 197, 198, 209, 213, 214, 215, 216, 224, 227, 231, 232, 247, 238, 240, 252, 255, 261, 262, 264, 265, 267, 274, 275, 277, 281, 287, 289, 297, 304, 306, 311, 313, 316, 318, 319, 329, 330, 331, 335, 336
Lilium auratum 27
—candidum 27
—pardalinum 27
—speciosum 27
Lily 265
—African 162
—Blackberry 185
Lily-of-the-Valley 33
Lily propagation 27, 28
—Thorn 203
—tiger 30
—Triplet 329
Lime, Spanish 273
Limits of graftage 77
Linaceæ 232, 266, 309
Linkia 295
Lion's Ear 264
—Foot 264
—Tail 264
Loasaceæ 188, 267, 273
Lobeliaceæ 228, 267
Locust seeds, treatment of . 17, 18
Lodeman, quoted . 111, 117, 148, 184
Loganiaceæ 239, 267
Loosestrife 269
Loranthaceæ 332
Love-in-a-Mist 279
Lungwort 273
Lupine 268
Lycopodiaceæ 269, 317
Lycopodiums 21
Lycopersicum esculentum . . . 328
Lythraceæ . 220, 262, 263, 269, 304
Madwort 165
Magnoliaceæ 205, 228, 257, 260, 267, 269, 273, 316
Mahaleb Cherry 147

Page
Mahalebs 206
Mallet cuttings 55
—for grafting 126
Mallow, Globe 320
—Poppy 197
Malpighiaceæ 194, 270, 322
Malvaceæ 158, 165, 189, 197, 217, 231, 252, 254, 270, 280, 320
Mammee Apple 270
Man-and-Wife 32
Management of nurseries . . 138
Manetti Rose 96
Mangifera Indica 271
Manna tree 163
Manuring nursery lands . . . 139
Maple cuttings 58
Maples, grafting 115
Marigold, Pot 196
Marrubium vulgare 254
Marvel of Peru 274
Matrimony Vine 269
May-apple 33
Mazzards 206
Medick 272
Melastomaceæ 176, 186, 190, 221, 263, 272, 320
Meliaceæ 203, 273, 315
Melissa officinalis 181
Menispermaceæ . . . 211, 213, 273
Mentha piperita 294
—Pulegium 294
—viridis 320
Mespilus Germanica 272
Mice, to protect from 146
Micropiper 294
Milk Vetch 179
Milkwort 304
Mock Privet 296
Moisture, for seeds 1
Mold in cellars 145
Monkey-flower 274
Moonflower seeds 18
Morellos 207
Morren, quoted 77
Morus alba 111, 275

Page
Morus Japonica 90
—nigra 275
—rubra 90, 111, 275
Moss for seeds 20, 21
Mottet, mentioned 23
Mound layering 39
Mountain Ash for pear stock . 74
— —seeds 17
—Ebony 183
—Fringe 161
—Holly 278
Mulberry 90, 111, 116
—French 196
—Paper 191
Musa paradisiaca 182
—Sapientum 182
Musk Plant 274
Muslin for tying 103
—screens 6
Mustard, Hedge 232
Myconia 308
Myrica asplenifolia 215
Myricaceæ 115, 277
Myristicaceæ 277
Myrobalan plum 148
—stocks 174. 291, 301
Myrrh 277
Myrsinaceæ 175
Myrtaceæ 181, 183, 197, 198, 200, 201, 232, 249, 272, 277
Myrtle 277, 332
Naiadaceæ 170, 287
Nasturtium Armoracia 254
—officinale 333
Natural graft 82
Navelwort 217
Nelumbium 278
—seeds 18
Nepenthaceæ 278
Nepeta Cataria 203
Nettle, Stingless 298
Neumann's cutting-pot 51
N. Y. Exp. Sta., quoted (note) . 139
N. Y. Exp. Sta., seed testing . 11
New Zealand Flax 297

Page
Nitrate of Soda 142
Nitrogen in lands 139, 142
Norway Spruce stocks . . 157, 298
Nursery lands 139
Nut-trees, cuttings 58
Nyctaginaceæ 158, 189, 274,
Nymphæaceæ . . 194, 278, 280, 332
Oakesia 330
Oaks, cuttings of 58
Ochnaceæ 280
Ocymum Basilicum 183
—minimum 183
Offsets 32
Olacineæ 280
Oleaceæ 209, 236, 237,
259, 265, 281, 287, 296, 325
Oleander 65, 66, 67
Olibanum tree 189
Olive, knaurs on 64
—Wild 229
Onagraceæ 212, 237, 280
Onion, top 30
—seed tests 10
Orchidaceæ 158, 159, 160,
161, 162, 168, 170, 182, 183, 188, 190,
191, 193, 195, 198, 203, 215, 222, 225,
227, 230, 262, 268, 272, 274, 280, 281,
284, 296, 315, 321, 327, 328, 331, 336
Orchids 60
—seeds 20
Otaheite Apple or Plum . . . 321
—Gooseberry 210
Othonna 287
Own-rooted trees . . . 87, 110, 153
Palay 334
Palmaceæ 159, 175,
179, 181, 189, 190, 195,
201, 205, 214, 224, 227, 233 239,
259, 261, 265, 267, 271, 288, 298, 300,
305, 306, 308, 313, 327, 328, 331, 333
Pandanaceæ 237, 289
Papaveraceæ 175, 188,
230, 232, 272, 289, 311, 315
Papver seeds 8
Paper Mulberry 191

Page
Paradise stock 73, 74, 148
— —layering 39
Parnassus, Grass of 289
Parrot Beak 213
Parsley, Cow 252
Parsnip, Cow 252
Passiflora edulis 242
Passifloraceæ 200, 242, 290
Passion Flower 290
Pastinaca sativa 289
Pea, Glory 213
—seed tests 10
Peach, stocks for 74
Peaches, budding 103
Peach-pits, treatment of . . . 17
Peach root cuttings 61
Pear, Alligator or Avocado . . 295
—root cuttings 61
—seeds, importing 19
Pears, dwarf 74, 133
—manuring 140, 141
—stocks for 75, 76, 77
Pea-tree 318
— —Siberian 200
Pedalineæ 271, 318
Pegging down 37
Peninsula Hort. Soc., quoted . 82
Peony, grafting 128, 129
Pereskia species, stocks . . 231
Pfeiffera 310
Phænocodon 262
Phalangium 169
Phœnix dactylifera 224
Phœnicophorum 179
Phosphoric acid in lands . . . 139
Phyllocereus 297
Phytolaccaceæ 298, 311
Picotia 281
Piece-root-grafting 149
Pieron, mentioned 131
Pilophora 271
Pin-cushion Flower 316
Pine 299
—dwarfing 148
Pine-apple 60

Page

Piperaceæ 294, 299
Pipping 171
Pips 33
Pisum sativum 290
Pitch for waxes 136
Pittosporaceæ 186, 300
Planer-tree 300
Plane-tree, knaur 64
Plantaginaceæ 300
Plantain Lily 238
Plastics 134
Platanaceæ 300
Plate-budding 105
Platyzamia 227
Plum, Coco 210
—dwarfing 148
Plumbaginaceæ . . . 176, 303, 322
Plums for peach stocks . . . 74
—manuring 140, 141
Podophyllum 33
Poiretia 255
Poke 298
Polemoniaceæ 200, 213, 267, 297, 303
Polygalaceæ 262, 304
Polygonaceæ 213, 275, 304, 310, 313, 320
Polygonum Sachalinense . . 313
Polypody 304
Pomme Blanche 306
Pontederia azurea 229
Pontederiaceæ 229, 304
Poplar 304
Poppy, California 232
—mallow 197
Porrum 164
Portugal Crakeberry 216
Portulacaceæ 195, 265, 304
Potash in lands 139
—to clean seeds 18
Potato 32
Potatoes, cuttings 59, 60
Pot-layering 40
—Marigold 196
Prairie Clover 295

Page

Preparation of seeds 15
Prickly Comfrey 62
Primrose 305
—Cape 323
Primula Auricula 180
Primulaceæ 167, 180, 217, 221, 228, 254, 269, 305, 319
Primula seeds 21
Privet, Mock 296
Prong-budding 103
Propagating-frames 45
Prosartes 227
Proteaceæ . . 182, 248, 250, 261, 295
Pruning trees 146
Prunus Amygdalus 164
—Americana 76
—Armeniaca 174
—Avium 206
—Besseyi 207
—Cerasus 206
—dasycarpa 174
—domestica 76
—Japonica 164
—Mahaleb 206
—Mume 174
—Pennsylvanica 207
—Persica 290
—Pissardi 76
—pumila 207
—Simonii 303
—species 300
Pseudo bulbs 60
Ptarmica 159
Puccoon, Red 315
Punica Granatum 304
Purslane 304
Putty-Root 170
Pyrus Cathayensis 307
—communis 291
Pyrus Cydonia 307
—Germanica 272
—Japonica 307
—Malus 170
—Sinensis 291
Quaking Grass 160

Page
Quince . . 62, 73, 74, 77, 78, 97, 133
Quinces, layering 39
Quince stocks 292
Raffia 101
Ragweed 317
Ragwort 287
Ranunculaceæ 160, 161, 168, 174, 179, 198, 211, 212, 215, 225, 231, 252, 279, 288, 308, 326, 329, 335
Ranunculus, Globe 329
Raphanus sativus 308
Raphia Ruffia 101
Rapunculus 298
Raspberry 35
Rattle-box 219
Recipes for wax 134
Red Puccoon 315
Redwood 317
Reed 177
—Mace 330
Regermination 9
Resedaceæ 309
Resting of land 140, 141
Retinospora cuttings 64
Rhacoma 276
Rhamnaceæ 186, 203, 255, 260, 288, 310
Rhizomes 32
R. I. Exp. Sta., seed testing . . 14
Rhododendron, grafting . . . 115
—seeds 20
Rhubarb 33
—cuttings 61
Ribes aureum 220
—Grossularia 241
—nigrum 220
—oxyacanthoides 241
—rubrum 220
Rice, Indian or Wild 336
Ring-budding 106
Ringing layers 38
Roberts, quoted 139
Rock Cress 174
Rock-foil 316
Root cuttings 57, 60

Page
Root-grafted trees 148
—grafting 107, 109, 132, 148
Rooting of trees 150
Root-tip 35, 36
Roots on cuttings 55
Rootstocks 32
Rosaceæ 161, 163, 164, 166, 170, 174, 187, 193, 206, 210, 217, 218, 225, 233, 240, 260, 261, 266, 272, 280, 290, 291, 297, 300, 305, 306, 307, 308, 312, 313, 321, 322, 333, 334
Rose 312
—Acacia 311
—Bay 310
—Moss 304
—seeds, treatment of 17
—stock 96
Roses 33
—cuttings 67, 68, 69
—graftage 76
Rosin-plant 318
Rosmarinus officinalis 313
Rowell, William M., quoted . 283
Rubiaceæ 161, 178, 189, 190, 193, 203, 209, 211, 214, 219, 238, 250, 253, 255, 258, 268, 270, 279, 298
Rubus Canadensis 225
—occidentalis 308
—phœnicolasius 334
—strigosus 308
—trivialis 225
—villosus 187
—vitifolius 225
Rumex 320
Runners 35
Rush, flowering 193
Rutaceæ 161, 162, 179, 189, 197, 209, 212, 226, 231, 241, 262, 264, 266, 282, 304, 306, 313, 335
Saccharum officinarum 324
Saccolabium 161, 296
Sacred Bean 278
Saddle-grafting 113
Saghalin 313
Salicaceæ 304, 314

Page

Sallow 314
—Thorn 253
Salmia 315
Salsify, Black 316
Salt-tree 250
Salvia officinalis 314
Sandal-tree 315
Sand Myrtle 264
—Verbena 158
Sandwort 175
Sapindaceæ 159, 162, 262, 273, 278, 321, 335
Sapotaceæ 210, 258, 315
Sarcocarpon 260
Sarcogonum 275
Sarraceniaceæ 224, 315
Satin Flower 319
Satureia hortensis 315
—montana 315
Saxifragaceæ 160, 178, 220, 224, 225, 237, 241, 252, 256, 258, 289, 296, 311, 316, 327
Saxifrage 316
Scalding seeds 17
Scandix cerefolium 208
Scheeria 159
Scitamineæ 182, 196, 199, 217, 220, 251, 271, 275, 323, 336
Schæfell's healing paint . . . 138
Schœnoprasum 164
Sclarea 314
Screens, for seeds 5
Scrophulariaceæ 165, 168, 169, 191, 196, 205, 226, 231, 239, 266, 272, 274, 290, 294, 313, 314, 328, 331
Sea Buckthorn 253
—Lavender 322
—Pink 176, 322
Seedage 1
Seeds 1–25
—aquatic 21
—cleaning 18
—depth to sow 21
—light on 8
—moisture for 1

Page

Seeds, moss for 2-, 21
—scalding 17
—soil for 20
—sowing 20
—stratifying 15
—temperature for 7
—testing 9
—transporting 19
Seed-grafting 131
Selaginellas 24
Separation 26
Serangium 274
Serpentine layering 37
Shade for seeds 5, 6, 7
Shield-budding 95
—grafting 116
Shin-leaf 306
Shed screens 6, 7
Siberian Crab 75, 76
—Pea-tree 200
Side-graft 115, 116
Silkweed 177
Silver fir 157
—sand 54
—weed 176
Simarubaceæ 163, 191, 307
Sinapis species 276
Sium Sisarum 319
Skoke 298
Slat screens 6
Slip 65
Smoke Vine 161
Snake Gourd 329
—root, Button 265
Snowdrop Tree 250
Soaking seeds 2, 3, 16, 17
Soils, adapting to by graftage . 74
—for cuttings 54
Soil for seeds 20
Soils, management 139
Solanaceæ 179, 205, 222, 229, 269, 270, 279, 295, 296, 297, 304, 316, 319, 328
Solanum Melongena 229
—tuberosum 304

Page
Solomon's Seal, False 319
Southernwood 176, 334
Sowbread 221
Sowing seeds 20
Spanish Lime 273
Spawn 31
Spencer, Herbert, quoted . . . 86
Sphagnum for seeds 20
Spinacia oleracea 321
Spireas 33, 69, 70
Splice-grafting 113
Spoke 298
Spores, sowing 21, 24
Sprouting chamber 14
—cups 14
—of trees 147
Spruce, dwarfing 148
Spurge 233
St. Domingo Apricot 270
St. John's Bread 201
St. Peter's Wort 325
Stag's-Horn Fern 300
Starwort 178
Statistics of nurseries 94
Stem cuttings 62
—grafting 107
Sterculiaceæ 158, 261, 263 269, 322, 327
Stick of buds 97
Stingless Nettle 298
Storax 324
Storing of trees 143
Stork's Bill 294
Stramonium 224
Stratification 15
String for tying 100, 103
—waxed 137
Strawberry Tomato 297
Straw, for heeling-in 146
Stub 118
Styracaceæ 250, 324, 325
Suckers 60
Sulphate of ammonia 142
Sulphur for fungus 145
Sweet Fern 215

Page
Sweet Gale 277
—Gum 266
—Potato 32
— —cuttings 59
Sweet-scented Shrub 198
— —Verbena 165
Syringa (Philadelphus) 296
Taccaceæ 325
Tamarind 325
Tamariscineæ 326
Tamarisk 326
Tanacetum vulgare 326
Taraxicum officinale 223
Tar for wounds 138
Tear Tree 222
Telanthera Bettzichiana . . . 165
Temperature for seeds 7
Ternstrœmiaceæ 160, 198, 201, 242, 323, 326
Testing of seeds 9
Texas Exp. Sta., quoted . . . 105
Thistle, Globe 229
Thomas knife 120
Thorn, for pear stock 77
—seeds, treatment of 17, 18
Throatwort 328
Thuya cuttings 64
Thymelæaceæ . . 223, 227, 262, 323
Thymus vulgaris 327
Thyrse Flower 327
Tiger Flower 327
Tiliaceæ 248, 320, 328
Tip 35, 36
Toadflax 266
Tomato, cleaning seeds 18
—seed test 10
—Tree, of Jamaica 222
Tongueing 38
Tools for grafting 119, 120
Toothpicks for cuttings 67
Toothwort 225
Top-grafting 107, 122
Tornelia 274
Tracy planter 22
Tragopogon porrifolius 314

Page
Transportation of seeds . . . 19
Treed lands 140
Tree Tomato of Jamaica . . . 222
Trigonella Fœnum-Græcum . . 234
Trimming trees 146
Triplet Lily 329
Tropical Almond 326
Trumpet Leaf 315
Tsuga Canadensis 252
Tubers 32
Tuber cuttings 59
Tubular-budding 106
Tulip 329
Turmeric 220
Turnip, seed tests 10
Turtle-head 205
Twig-budding 105
Tying of buds 100
Typhaceæ 330
Umbelliferæ 160, 168, 193, 203, 208, 216, 219, 226, 232, 234, 252, 268, 277, 289, 319
Umbrella Pine 316
—tree 269
Unions, of grafted plants . . . 87
Upland Cress 182
Urticaceæ 177, 191, 200, 228, 235, 236, 255, 269, 275, 298, 300, 330
Vaccaria 315
Vaccinium macrocarpon . . . 218
Valerian 330
Valerianaceæ 216, 330
Valerianella 216
Variegation and graftage . . . 77
Variegations 62, 72
Veneer-budding 106
—grafting 113
Veneer-graft union 88
Verbena, Lemon 165
Verbenaceæ 165, 189, 196, 201, 212, 262, 266, 331, 333
Vermin, to protect from . . . 146
Vervain 331
Vetch, Bitter 287

Page
Vetch, Kidney 169
—Milk 179
Vetchling 263
Viburnum, dwarfs 148
—layer 36
—treatment of seeds 17, 18
Viburnums, cuttings of 58
Vicia Faba 334
—sativa 331
Violaceæ 257, 332
Violet 332
Virility of grafted plants . . . 90
Viscaria 268
Vitaceæ 166, 211, 242
Vochysiaceæ 333
Wake-Robin 329
Wall Cress 174
Walnut, grafting 131
Walnuts, treatment of 17
Water Bean 278
—Caltrops 328
—Chinquapin 278
—Lily seeds 21
—Plantain 163
—Platter 332
Waxberry 325
Waxes 134
Wax Myrtle 277
Waxing the wounds 122
Weeping trees 73
Weigela 67, 68
Wendlandia 213
Whip-graft union 89
—grafting 108
Whistle-budding 106
White Hellebore 331
Whitlow Grass 228
Whole-root-grafting 149
Wild Hyacinth 316
—Rice 336
Willow cuttings 55, 56
Window-garden plants . . . 65, 67
Wintera 228
Winter Aconite 231
—budding 105

	Page		Page
Winter Cherry	297	Yarrow	159
Wintergreen	238, 306	Yews from cuttings	64
Winter Olive	229	Youth-and Old-Age	336
Wolf's Bane	160	Yucca cuttings	65
Wounds, waxes for	137	Zea Mays	270
Woundwort	321	Zephyr Flower	336
Wrightia (Wallichia)	333	Zerumbet	336
Xylosteum	268	Zygophyllaceæ	248, 336

Styrax (Storax). *Styraceae.*

Seeds, which must be stratified, or else sown as soon as ripe. They usually lie dormant the first year. Also by layers and cuttings of green wood. Can be grafted upon other storaxes, or upon *Halesia tetraptera.*

Sugar Cane (*Saccharum officinarum*). *Gramineae.*

Cuttings of the stems. The cuttings should possess a node or joint which bears one or more good buds. These cuttings are planted directly in the field, and the plants will reach maturity in two or three months. Propagation by seeds has been supposed to be impossible, but recent experiments at Kew indicate that it can be done.

Sumach. See Rhus.

Sundew. See Drosera.

Sunflower. See Helianthus.

Sun Rose. See Helianthemum.

Swainsona. *Leguminosae.*

Seeds. Green cuttings under cover.

Swan River Daisy. See Brachycome.

Sweet Brier. See Rosa.

Sweet Cicely. See Myrrhis.

Sweet Pea. See Lathyrus.

Sweet Potato (*Ipomoea Batatas*). *Convolvulaceae.*

Sweet potato plants are grown in hotbeds, coldframes or forcing houses (depending upon the latitude) from sound tubers of medium size. The tuber is laid upon a sandy or other loose bed, and is then covered with sand or sandy loam to a depth of 1 or 2 inches. Sometimes, to guard against rot, the tubers are not covered until the sprouts begin to appear. The tubers may be laid thickly upon the bed, but they are less apt to rot if they do not touch each other. Sometimes the tubers are cut in two lengthwise, the cut surface being placed down, in order to place all the plant-giving surface uppermost. In four or five weeks the young plants—3 to 5 inches high—are pulled off and planted, and others soon arise to take their places. One hand should be held firmly upon the soil over the tuber while the sprout is pulled off, to keep it in place. Three or four crops of sprouts may be obtained from each tuber.